Ho Math Chess　何数棋谜　妈!我会棋谜式
Mom! I Learn Multiplication Using Math-Chess-Puzzles
Student's Name _____ Date _____
2007 - 2017 © Frank Ho, Amanda Ho, All rights reserved.　www.homathchess.com

Mom! I Learn Multiplication Using Math-Chess-Puzzles Connection

棋谜式乘法

Frank Ho   Amanda Ho

何数棋谜 培训

Ho Math Chess Learning Centre

Table of Contents

Ho Math Chess　　何数棋谜　妈！我会棋谜式乘法啦！
Mom! I Learn Multiplication Using Math-Chess-Puzzles Connection!

Student's Name _____ Date _____

2007 - 2017 © Frank Ho, Amanda Ho, All rights reserved.　　www.homathchess.com

Ho Math Chess　何数棋谜　妈！我会棋谜式乘法啦！

Mom! I Learn Multiplication Using Math-Chess-Puzzles Connection!

Student's Name _____ Date _____

2007 - 2017 © Frank Ho, Amanda Ho, All rights reserved.　　www.homathchess.com

2007 - 2017 © Frank Ho, Amanda Ho, All rights reserved.   www.homathchess.com

Preface 2017

I have added one digit divided by one digit division in this new updated multiplication workbook. The reason is that it does not matter how we train math teachers to ask students every time before the class starts on what they are learning at day school math class, sometimes the teacher forgot or the students were not able to communicate well, so one thing happened was the students already were doing 3-digit times 2-digit by using this workbook, but they have never done any divisions while their day schools start to do one-digit divided by one-digit. We do not want the students to carry too many workbooks, so this division addition serves as a transition before purchasing a division workbook.

Ho Math Chess   何数棋谜   妈！我会棋谜式乘法啦！
Mom! I Learn Multiplication Using Math-Chess-Puzzles Connection!

Student's Name _____ Date _____

2007 - 2017 © Frank Ho, Amanda Ho, All rights reserved.   www.homathchess.com

About Ho Math Chess™ Math Workbooks

I have taught students from grade 1 to grade 12 since I opened the Vancouver *Ho Math Chess* Learning Centre in 1995. I have personally witnessed on how some students suffered because they could not master some very basic computational skills. I do not want to create a workbook, which is about practice, practice, and more practice of computational skills. This has motivated me to create a workbook that would be very different from the conventional ones in terms of the way the questions are presented to the students. I wanted students to learn basic computation skills by using the carefully designed worksheets so that students can master basic computation skills in an intuitive way. These worksheets were being designed while I actually watched student's work and modified accordingly to their responses.

I had an idea to create a computational workbook, which integrates chess knowledge, puzzles, and math in such a way that students could learn how to transfer abstract symbols into numerical values and then calculating the results by using puzzles-like problems. This idea may sound very simple but the result is much more profound – not only students learn to do math in multi-step, they also learn how to process information by converting abstract symbols into numerical values, which is important in learning critical thinking skill.

One very noticeable computation format in *Ho Math Chess* math workbooks is the way computation directions are presented - it is no longer just a linear fashion; instead, students work on computations in all kinds of directions: top down, bottom up, left to eight, right to left , diagonally, and even circular motion. For example, the multiplication workbook computation format is designed in such a way that it takes the boredom out by using the format of multi-direction computation and multi-concept learning. Students could be introduced division computation procedure while working on multiplication and even equivalent fractions but without realizing that they are actually working on advanced math concepts and mechanic computation procedure beyond their grade level. One other example is that the factoring procedure is introduced while students are working on multiplication. These many embedded computational procedures included in the elementary level of math workbook will benefit students when they go to higher grades.

2007 - 2017 © Frank Ho, Amanda Ho, All rights reserved.   www.homathchess.com

My idea of using multi-direction, multi-operation, multi-procedure, multi-concept learning style is the very distinct and innovative way of creating these workbooks. Students found them less boring and even willing to do the same worksheets the second time if they did not master the first time.

I am hoping by working through these addition, subtraction, multiplication workbooks, the division would be just a matter of fine-tuning its computation procedure.

In 2014, all computation workbooks have taken major upgrades to include truly math and chess integrated material, this idea is world first and these worksheets formats are also world first. With these releases of many new and innovative workbooks, the math teaching and tutoring has taken the entire math tutoring to a revolutionary stage. Because of the creation of integrated math, chess, and puzzles integrated workbooks, Ho Math Chess has made the dream of fun math teaching becomes true.

Students at Ho Math Chess have enjoyed math even more than the previous workbooks and we see dramatic changes in student's attitude, they are happier and more willing to work on math.

Frank Ho
Amanda Ho

July 2014

In 2015, we added a new part called intelligent worksheets which allow students to figure out the operator and this is an innovative idea in computation format because all other math workbooks on the market all pre-define their operators for students and students just calculate the results.

November 2015
Frank Ho
Amanda Ho

2007 - 2017 © Frank Ho, Amanda Ho, All rights reserved.   www.homathchess.com

Chess pieces and their mathematical values

Symbols of chess pieces	Names of chess pieces	Mathematical values
	Queen	9
	Rook	5
	Bishop	3
	Knight	3
	Pawn	1
	King	0

2007 - 2017 © Frank Ho, Amanda Ho, All rights reserved.　　www.homathchess.com

From addition to multiplication

Fill in each _____ with a mathematical expression.

Diagram	Explanation	Addition	Multiplication	Comments
♟♟ ♟♟ ♟♟	3 groups of ♟♟	2 + 2 + 2 = 6	3 × 2 = 6	**3 times 2 is** (equals) **6.** 3 multiplied by 2 is (equals) 6. The product of the factors 3 and 2 is (equals) 6. Three 2s equal 6. 3 sets of 2 horses are 6 horses. Factor 3 times factor 2 is product 6. Multiplicand 3 times multiplier 2 is product 6.
♟♟ ♟♟ ♟♟ ♟♟	4 groups of ♟♟.	2 + 2 + 2 + 2 = 8	_____	
♟♟♟♟ ♟♟♟♟	2 groups of ♟♟♟♟.	4 + 4 = ___	_____	
♟♟♟♟ ♟♟♟♟ ♟♟♟♟ ♟♟♟♟	4 groups of ♟♟♟♟	_____	_____	
♟♟♟ ♟♟♟ ♟♟♟ ♟♟♟	4 groups of 3 ♟♟♟	_____	_____	

2007 - 2017 © Frank Ho, Amanda Ho, All rights reserved.　www.homathchess.com

Multiplication table

×	0	1	2	3	4	5	6	7	8	9
1	0	1	2	3	4	5	6	7	8	9
2	0	2	4	6	8	10	12	14	16	18
3	0	3	6	9	12	15	18	21	24	27
4	0	4	8	12	16	20	24	28	32	36
5	0	5	10	15	20	25	30	35	40	45
6	0	6	12	18	24	30	36	42	48	54
7	0	7	14	21	28	35	42	49	56	63
8	0	8	16	24	32	40	48	56	64	72
9	0	9	18	27	36	45	54	63	72	81

Ho Math Chess　何数棋谜　妈！我会棋谜式乘法啦！

Mom! I Learn Multiplication Using Math-Chess-Puzzles Connection!

Student's Name _____ Date _____

2007 - 2017 © Frank Ho, Amanda Ho, All rights reserved.　　www.homathchess.com

Multiplying with 0 and 1

2 X 1 2▢	♙ X 3 3▢	1 X 4 4▢	0 X ♗ 0▢	2 X ♛ 0▢
5 X ♙ 5▢	1 X 5 5▢	7 X ♙ 7▢	9 X 1 9▢	8 X 1 8▢
2 X ♚ 0▢	9 X 0 0▢	♚ X 2 0▢	1 X 3 3▢	2 X 0 0▢
♙ X 7 7▢	8 X 1 8▢	4 X 0 0▢	1 X 8 8▢	♚ X 5 0▢

Ho Math Chess 何数棋谜 妈！我会棋谜式乘法啦！
Mom! I Learn Multiplication Using Math-Chess-Puzzles Connection!
Student's Name _____ Date _____
2007 - 2017 © Frank Ho, Amanda Ho, All rights reserved.   www.homathchess.com

Counting 2's multiples (Doubling)

Place an O over every multiple of 2.

1	2	3	4	5	6	7	8	9	10
11	12	13	14	15	16	17	18	19	20
21	22	23	24	25	26	27	28	29	30
31	32	33	34	35	36	37	38	39	40
41	42	43	44	45	46	47	48	49	50
51	52	53	54	55	56	57	58	59	60
61	62	63	64	65	66	67	68	69	70
71	72	73	74	75	76	77	78	79	80
81	82	83	84	85	86	87	88	89	90
91	92	93	94	95	96	97	98	99	100

What are the unit digits of 2's multiples ? _____ **2, 4, 6, 8, 0**_____.

Counting 2's multiples

Circle the following 2's multiples.

1 2 3 4 5 6 7 8 9 10 11 12 13 14 15 16 17 18

Fill in the following ☐ with a number.

Sequence	1	2	3	4	5	6	7	8	9
Add 2	☐	4	6	☐	10	12	☐	☐	18

Sequence	♙	2	♝	4	♜	6	7	8	♛
Add 2	2	☐	6	☐	10	☐	☐	☐	18

Sequence	1	2	3	4	5	6	7	8	9
Add 2	2	☐	6	☐	10	☐	14	☐	18

Sequence	♙	2	♝	4	♜	6	7	8	♛
Add 2	☐	4	☐	8	☐	12	☐	16	☐

Sequence	1	2	3	4	5	6	7	8	9
Add 2	☐	☐	☐	8	☐	12	☐	16	☐

Sequence	♙	2	♝	4	♜	6	7	8	♛
Add 2	☐	4	☐	☐	☐	12	☐	16	☐

2007 - 2017 © Frank Ho, Amanda Ho, All rights reserved.   www.homathchess.com

2 times

Fill in each _____ with multiplication expression.

$2 \times 1 =$	♟ ♟ =	$= 1 \times 2 =$	♟ ♟
$2 \times 2 =$	♟ ♟ ♟ ♟ =		♟ ♟ ♟ ♟
$2 \times 3 =$	♟ ♟ ♟ ♟ ♟ ♟ =	$=$ _____ $=$	♟ ♟ ♟ ♟ ♟ ♟
$2 \times 4 =$	♟ ♟ ♟ ♟ ♟ ♟ ♟ ♟ =	$= 4 \times 2 =$	♟ ♟ ♟ ♟ ♟ ♟ ♟ ♟
$2 \times 5 =$	♟ ♟ ♟ ♟ ♟ ♟ ♟ ♟ ♟ ♟ =	$=$ _____ $=$	♟ ♟ ♟ ♟ ♟ ♟ ♟ ♟ ♟ ♟
$2 \times 6 =$	♟ ♟ ♟ ♟ ♟ ♟ ♟ ♟ ♟ ♟ ♟ ♟ =	$= 6 \times 2 =$	♟ ♟ ♟ ♟ ♟ ♟ ♟ ♟ ♟ ♟ ♟ ♟
$2 \times 7 =$	♟ ♟ ♟ ♟ ♟ ♟ ♟ ♟ ♟ ♟ ♟ ♟ ♟ ♟ =	$=$ _____ $=$	♟ ♟ ♟ ♟ ♟ ♟ ♟ ♟ ♟ ♟ ♟ ♟ ♟ ♟
$2 \times 8 =$	♟ ♟ ♟ ♟ ♟ ♟ ♟ ♟ ♟ ♟ ♟ ♟ ♟ ♟ ♟ ♟ =	$= 8 \times 2 =$	♟ ♟ ♟ ♟ ♟ ♟ ♟ ♟ ♟ ♟ ♟ ♟ ♟ ♟ ♟ ♟

2007 - 2017 © Frank Ho, Amanda Ho, All rights reserved.　www.homathchess.com

2 times

$1 + 1 =$ ☐	$= 2 \times 1 =$	$1 \times 2 =$ ☐		♙ $+$ ♙ ☐ $= 2 \times 1 =$ ☐
$2 + 2 =$ ☐	$= 2 \times 2 =$	$2 \times 2 =$ ☐		2 $+ 2$ ☐ $= 2 \times 2 =$ ☐
$3 + 3 =$ ☐	$= 2 \times 3 =$	$3 \times 2 =$ ☐		♞ $+$ ♞ ☐ $= 2 \times 3 =$ ☐
$4 + 4 =$ ☐	$= 2 \times 4 =$	$4 \times 2 =$ ☐		4 $+ 4$ ☐ $= 2 \times 4 =$ ☐

Ho Math Chess　何数棋谜　妈！我会棋谜式乘法啦！
Mom! I Learn Multiplication Using Math-Chess-Puzzles Connection!

Student's Name _____ Date _____

2007 - 2017 © Frank Ho, Amanda Ho, All rights reserved.　　www.homathchess.com

$5 + 5 =$	☐	$= 2 \times$ ♖ $=$	$5 \times 2 =$	☐	$+$ ♖ / ☐ $= 2 \times 5 =$ ☐
$6 + 6 =$	☐	$= 2 \times 6 =$	$6 \times 2 =$	☐	6 $+ 6$ / ☐ $= 2 \times 6 =$ ☐
$7 + 7 =$	☐	$= 2 \times 7 =$	$7 \times 2 =$	☐	7 $+ 7$ / ☐ $= 2 \times 7 =$ ☐
$8 + 8 =$	☐	$= 2 \times 8 =$	$8 \times 2 =$	☐	8 $+ 8$ / ☐ $= 2 \times 8 =$ ☐
$9 + 9 =$	☐	$= 2 \times 9 =$	$9 \times 2 =$	☐	♕ $+$ ♕ / ☐ $= 2 \times 9 =$ ☐

2007 - 2017 © Frank Ho, Amanda Ho, All rights reserved.   www.homathchess.com

2 times

2 × 1 = ☐	Two times one is ☐	1 × 2 = ☐	One times two is ☐
2 × 2 = ☐	Two times two is ☐	2 × 2 = ☐	Two times two is ☐
2 × 3 = ☐	Two times three is ☐	3 × 2 = ☐	Three times two is ☐
2 × 4 = ☐	Two times four is ☐	4 × 2 = ☐	Four times two is ☐
2 × 5 = ☐	Two times five is ☐	5 × 2 = ☐	Five times two is ☐
2 × 6 = ☐	Two times six is ☐	6 × 2 = ☐	Six times two is ☐
2 × 7 = ☐	Two times seven is ☐	7 × 2 = ☐	Seven times two is ☐
2 × 8 = ☐	Two times eight is ☐	8 × 2 = ☐	Eight times two is ☐
2 × 9 = ☐	Two times nine is ☐	9 × 2 = ☐	Nine times two is ☐

2	♙	2	2	2
X 1	X 2	X 2	X 3	X ♗
☐	☐	☐	☐	☐

5	2	7	2	9
X 2	X 6	X 2	X 8	X 2
☐☐	☐☐	☐☐	☐☐	☐☐

7	2	♖	2	2
X 2	X 8	X 2	X 6	X 9
☐☐	☐☐	☐☐	☐☐	☐☐

2007 - 2017 © Frank Ho, Amanda Ho, All rights reserved.   www.homathchess.com

2	♙	4	2	2
X 1	X 2	X 2	X 4	X 3

2	♘	2	8	2
X 3	X 2	X 7	X 2	X 9

4	2	5	2	7
X 2	X 4	X 2	X ♗	X 2

2	9	3	3	♗
X ♘	X 2	X 2	X 2	X 2

2	2	2	2	2
X 3	X 6	X 7	X 3	X 9

2007 - 2017 © Frank Ho, Amanda Ho, All rights reserved.   www.homathchess.com

Oral practice

two one two	2 ♙ ☐	$\begin{array}{r} 1\ 1 \\ \times\quad 2 \\ \hline 22\ ☐\ ☐ \end{array}$
two two four	2 2 ☐	$\begin{array}{r} 2\ 2 \\ \times\quad 2 \\ \hline 44\ ☐\ ☐ \end{array}$
two three six	2 ♗ ☐	$\begin{array}{r} 3\ 3 \\ \times\quad 2 \\ \hline 66\ ☐\ ☐ \end{array}$
two four eight	2 4 ☐	$\begin{array}{r} 4\ 4 \\ \times\quad 2 \\ \hline 88\ ☐\ ☐ \end{array}$
two five ten	2 5 ☐	$\begin{array}{r} {}^{1}\,\\ 5\ 5 \\ \times\quad 2 \\ \hline 110\ ☐\ ☐\ ☐ \end{array}$

2007 - 2017 © Frank Ho, Amanda Ho, All rights reserved.　www.homathchess.com

Oral practice

two six twelve	2 6 ☐	$\begin{array}{r} {}^{1} \\ 6\,6 \\ \times2 \\ \hline 132\,☐☐☐ \end{array}$
two seven fourteen	2 7 ☐	$\begin{array}{r} 7\,7 \\ \times2 \\ \hline 154\,☐☐☐ \end{array}$
two eight sixteen	2 8 ☐	$\begin{array}{r} 8\,8 \\ \times2 \\ \hline 176\,☐☐☐ \end{array}$
two nine eighteen	2 9 ☐	$\begin{array}{r} 9\,9 \\ \times2 \\ \hline 198\,☐☐☐ \end{array}$
two five ten	2 ♖ ☐	$\begin{array}{r} 5\,5 \\ \times2 \\ \hline 110\,☐☐☐ \end{array}$
two six twelve	2 6 ☐	$\begin{array}{r} 6\,6 \\ \times2 \\ \hline 132\,☐☐☐ \end{array}$

2007 - 2017 © Frank Ho, Amanda Ho, All rights reserved.   www.homathchess.com

Fill in _____ and ☐ with answers.

Times	Grouping	Addition
2 × \$1 = ☐	2 of \$1 = \$2	\$1 + \$1 = ☐
♟ × \$2 = ☐	1 of \$2 = \$2	\$2
2 × \$2 = ☐	2 of ☐ = \$4	\$2 + \$2 = ☐

Fill in _____ and ☐ with answers.

Expression	Grouping	Addition
2 × \$1 =	2 of ☐ = \$2	\$♟ + \$♟ = ☐
1 × \$2 =	♟ of ☐ = \$2	\$2 + \$2 = ☐

2 × 1 = ☐ = ♟ × 2 = ☐	2 × 2 = ☐ = 2 × 2 = ☐
2 × ☐ = 2 = 1 × ☐ = ☐	2 × ☐ = 4 = 2 × ☐ = ☐

2 ♟ ☐	2 5 ☐	2 ♛ ☐	2 4 ☐	2 8 ☐
2 2 ☐	2 6 ☐	2 1 ☐	2 5 ☐	2 9 ☐
2 3 ☐	2 7 ☐	2 2 ☐	2 6 ☐	2 ♟ ☐
2 4 ☐	2 8 ☐	2 3 ☐	2 7 ☐	2 2 ☐

Student's Name _____ Date _____

2007 - 2017 © Frank Ho, Amanda Ho, All rights reserved.　www.homathchess.com

Fill in _____ and ☐ with answers.

Times	Grouping	Addition
$2 \times \$3 = \boxed{}$	$2 \text{ of } \boxed{} = 6$	$\$3 + \$3 = \boxed{}$
$3 \times \$2 = \boxed{}$	$3 \text{ of } \boxed{} = 6$	$\$2 + \$2 + \$2 = \boxed{}$

Fill in _____ and ☐ with answers.

Expression	Grouping	Addition
$2 \times \$4$	$2 \text{ of } \boxed{} = \$8$	$\$4 + \$4 = \boxed{}$
$4 \times \$2$	$4 \text{ of } \boxed{} = \$8$	$\$2 + \$2 + \$2 + \$2 = \boxed{}$

$2 \times 3 = \boxed{} = 3 \times 2 = \boxed{}$	$2 \times 4 = \boxed{} = 4 \times 2 = \boxed{}$
$2 \times \boxed{} = 6 = 3 \times \boxed{} = \boxed{}$	$2 \times \boxed{} = 8 = 4 \times \boxed{} = \boxed{}$

$2 \, ♙ \, \boxed{}$	$2 \, 5 \, \boxed{}$	$2 \, 9 \, \boxed{}$	$2 \, 4 \, \boxed{}$	$2 \, 8 \, \boxed{}$
$2 \, 2 \, \boxed{}$	$2 \, 6 \, \boxed{}$	$2 \, 1 \, \boxed{}$	$2 \, ♖ \, \boxed{}$	$2 \, ♕ \, \boxed{}$
$2 \, 3 \, \boxed{}$	$2 \, 7 \, \boxed{}$	$2 \, 2 \, \boxed{}$	$2 \, 6 \, \boxed{}$	$2 \, 1 \, \boxed{}$
$2 \, 4 \, \boxed{}$	$2 \, 8 \, \boxed{}$	$2 \, 3 \, \boxed{}$	$2 \, 7 \, \boxed{}$	$2 \, 2 \, \boxed{}$

2007 - 2017 © Frank Ho, Amanda Ho, All rights reserved.   www.homathchess.com

Fill in _____ and ☐ with answers.

Times	Grouping	Addition
2 × \$5 = ☐	2 of ☐ = 10	\$5 + \$ ♖ = ☐
5 × \$2 = ☐	5 of ☐ = 10	\$2 + \$2 + \$2 + \$2 + \$2 = ☐

Times	Grouping	Addition
2 × \$6 = ☐	2 of ☐ = \$12	\$6 + \$6 = ☐
6 × \$2 = ☐	6 of ☐ = \$12	\$2 + \$2 + \$2 + \$2 + \$2 + \$2 = ☐

2 × ♖ = ☐ = 5 × 2 = ☐	2 × 6 = ☐ = 6 × 2 = ☐
2 × ☐ = 10 = 5 × ☐ = ☐	2 × ☐ = 12 = 6 × ☐ = ☐

2 ♙ ☐	2 5 ☐	2 ♕ ☐	2 4 ☐	2 8 ☐
2 2 ☐	2 6 ☐	2 1 ☐	2 ♖ ☐	2 9 ☐
2 3 ☐	2 7 ☐	2 2 ☐	2 6 ☐	2 ♙ ☐
2 4 ☐	2 8 ☐	2 3 ☐	2 7 ☐	2 2 ☐

Ho Math Chess 何数棋谜 妈！我会棋谜式乘法啦！
Mom! I Learn Multiplication Using Math-Chess-Puzzles Connection!

Student's Name _____ Date _____

2007 - 2017 © Frank Ho, Amanda Ho, All rights reserved.   www.homathchess.com

Fill in _____ and □ with answers.

Times	Grouping	Addition
2 × $7 = □	2 of □ = 14	$7 + $ 7 = □
7 × $2 = □	7 of □ = 14	$2 + $2 + $2 + $2 + $2 + $2 + $2 = □

Times	Grouping	Addition
2 × $8 = □	2 of □ = $16	$8 + $8 = □
8 × $2 = □	8 of □ = $16	$2 + $2 + $2 + $2 + $2 + $2 + $2 + $2 = □

2 × 7 = □ = 7 × 2 = □	2 × 8 = □ = 8 × 2 = □
2 × □ = 14 = 7 × □ = □	2 × □ = 16 = 8 × □ = □

2 1 □	2 5 □	2 ♛ □	2 4 □	2 8 □
2 2 □	2 6 □	2 1 □	2 5 □	2 9 □
2 ♙ □	2 7 □	2 2 □	2 6 □	2 1 □
2 4 □	2 8 □	2 3 □	2 7 □	2 2 □

Ho Math Chess 何数棋谜 妈！我会棋谜式乘法啦！
Mom! I Learn Multiplication Using Math-Chess-Puzzles Connection!

Student's Name _____ Date _____

2007 - 2017 © Frank Ho, Amanda Ho, All rights reserved.   www.homathchess.com

Fill in _____ and ☐ with answers.

Times	Grouping	Addition
2 × $9 = ☐	2 of ☐ = 18	$♛ + $♛ = ☐
7 × $2 = ☐	7 of ☐ = 14	$2 + $2 + $2 + $2 + $2 + $2 + $2 = ☐

Times	Grouping	Addition
2 × $8 = ☐	2 of ☐ = $16	$8 + $8 = ☐
8 × $2 = ☐	8 of ☐ = $16	$2 + $2 + $2 + $2 + $2 + $2 + $2 + $2 = ☐

2 × 9 = ☐ = ♛ × 2 = ☐	2 × 8 = ☐ = 8 × 2 = ☐
2 × ☐ = 18 = 9 × ☐ = ☐	2 × ☐ = 16 = 8 × ☐ = ☐

2 1 ☐	2 5 ☐	2 9 ☐	2 4 ☐	2 8 ☐
2 2 ☐	2 6 ☐	2 ♙ ☐	2 ♖ ☐	2 ♛ ☐
2 ♞ ☐	2 7 ☐	2 2 ☐	2 6 ☐	2 1 ☐
2 4 ☐	2 8 ☐	2 ♞ ☐	2 7 ☐	2 2 ☐

2007 - 2017 © Frank Ho, Amanda Ho, All rights reserved.   www.homathchess.com

Preparing for division

☐	☐	☐	☐	☐
X 2	X 2	X 4	X 2	X 3
4	6	8	10	12

☐	☐	☐	☐	☐
X 2	X 6	X 2	X 3	X 2
6	12	14	6	18

☐	☐	☐	☐	☐
X 4	X 2	X ♜	X 2	X 7
8	16	10	8	14

☐	☐	☐	☐	☐
X 2	X 2	X 2	X ♝	X 2
6	18	8	6	16

☐	☐	☐	☐	☐
X 4	X 5	X 2	X 2	X 7
8	10	16	8	14

Preparing for division

☐ X 2 = 2	X ☐ 2)2	☐)2 X 2
☐ X 2 = 4	X ☐ 2)4	☐)4 X 2
☐ X 2 = 6	X ☐ 2)6	☐)6 X 2
☐ X 2 = 8	X ☐ 2)8	☐)8 X 2
☐ X 2 = 10	X ☐ 2)10	☐)10 X 2
☐ X 2 = 12	X ☐ 2)12	☐)12 X 2
☐ X 2 = 14	X ☐ 2)14	☐)14 X 2

2007 - 2017 © Frank Ho, Amanda Ho, All rights reserved. www.homathchess.com

Preparing for division

☐ X 2 = 16	X ☐ 2)16	☐)16 X 2
☐ X 2 = 18	X ☐ 2)18	☐)18 X 2
☐ X 2 = 2	X ☐ 2)2	☐)2 X 2
☐ X 2 = 4	X ☐ 2)4	☐)4 X 2
☐ X 2 = 6	X ☐ 2)6	☐)6 X 2
☐ X 2 = 8	X ☐ 2)8	☐)8 X 2
☐ X 2 = 10	X ☐ 2)10	☐)10 X 2

Ho Math Chess   何数棋谜   妈!我会棋谜式乘法啦!
Mom! I Learn Multiplication Using Math-Chess-Puzzles Connection!

Student's Name _____ Date _____

2007 - 2017 © Frank Ho, Amanda Ho, All rights reserved.   www.homathchess.com

Cross multiplication

12 12 ↖ ↗ $\frac{6}{2} = \frac{6}{2}$	☐ **4** ↖ ↗ $\frac{2}{2} = \frac{2}{2}$	☐ **6** ↖ ↗ $\frac{2}{2} = \frac{3}{3}$	☐ **8** ↖ ↗ $\frac{2}{2} = \frac{4}{4}$
☐ **10** ↖ ↗ $\frac{2}{2} = \frac{5}{5}$	☐ **12** ↖ ↗ $\frac{2}{2} = \frac{6}{6}$	☐ **14** ↖ ↗ $\frac{2}{2} = \frac{7}{7}$	☐ **18** ↖ ↗ $\frac{2}{2} = \frac{9}{9}$
☐ **10** ↖ ↗ $\frac{2}{2} = \frac{5}{5}$	☐ **8** ↖ ↗ $\frac{2}{2} = \frac{4}{4}$	☐ **14** ↖ ↗ $\frac{2}{2} = \frac{7}{7}$	☐ **16** ↖ ↗ $\frac{2}{2} = \frac{8}{8}$
☐ **12** ↖ ↗ $\frac{2}{2} = \frac{6}{6}$	☐ **16** ↖ ↗ $\frac{2}{2} = \frac{8}{8}$	☐ **18** ↖ ↗ $\frac{2}{2} = \frac{9}{9}$	☐ **6** ↖ ↗ $\frac{2}{2} = \frac{3}{3}$

2007 - 2017 © Frank Ho, Amanda Ho, All rights reserved.　　www.homathchess.com

Different ways of writing multiplication (Learning division while doing multiplications)

$$\frac{4}{\Box\times}=2 \qquad \begin{array}{c}2\\ \times\\ 2\end{array} \qquad \frac{4}{\Box\times}=2$$

$$2\times\Box\quad =\quad \boxed{}\quad =\quad 2\times\Box$$

$$\begin{array}{r}\times\Box\\ 2\overline{)4}\end{array} \qquad \begin{array}{c}2\\ \times\\ \Box\end{array} \qquad \begin{array}{r}\times\Box\\ 2\overline{)4}\end{array}$$

$$\begin{array}{r}2\overline{)4}\\ \times\Box\end{array} \qquad\qquad \begin{array}{r}2\overline{)4}\\ \times\Box\end{array}$$

Ho Math Chess   何数棋谜   妈！我会棋谜式乘法啦！
Mom! I Learn Multiplication Using Math-Chess-Puzzles Connection!

Student's Name _____ Date _____

2007 - 2017 © Frank Ho, Amanda Ho, All rights reserved.   www.homathchess.com

Different ways of writing multiplication (Learning division while doing multiplications)

2007 - 2017 © Frank Ho, Amanda Ho, All rights reserved.　www.homathchess.com

Different ways of writing multiplication (Learning division while doing multiplications)

$$2$$
$$\times$$
$$4$$

$$\frac{8}{\boxed{}\times}=2 \qquad \frac{8}{\boxed{}\times}=4$$

$$\nwarrow \ || \ \nearrow$$

$$4 \times \boxed{} \ = \ \boxed{} \ = \ \boxed{} \times 4$$

$$\swarrow \ || \ \searrow$$

$$\times \boxed{} \qquad\qquad 4 \qquad\qquad \times \boxed{}$$

$$2\overline{)8} \qquad\qquad \times \qquad\qquad 4\overline{)8}$$

$$\boxed{}$$

$$|| \qquad\qquad\qquad\qquad ||$$

$$2\underline{)8} \qquad\qquad\qquad\qquad 4\underline{)8}$$
$$\times \boxed{} \qquad\qquad\qquad\qquad \times \boxed{}$$

32

2007 - 2017 © Frank Ho, Amanda Ho, All rights reserved.   www.homathchess.com

Different ways of writing multiplication (Learning division while doing multiplications)

Ho Math Chess   何数棋谜 妈！我会棋谜式乘法啦！
Mom! I Learn Multiplication Using Math-Chess-Puzzles Connection!
Student's Name _____ Date _____
2007 - 2017 © Frank Ho, Amanda Ho, All rights reserved.   www.homathchess.com

Different ways of writing multiplication (Learning division while doing multiplications)

$$2 \times 6$$

$$\frac{12}{\Box \times} = 2 \qquad \frac{12}{\Box \times} = 6$$

$$2 \times \Box \quad = \boxed{} = \quad 6 \times \Box$$

$$\begin{array}{r} \times\,\Box \\ 2\,\overline{)\,12} \end{array} \qquad \begin{array}{c} 6 \\ \times \\ \Box \end{array} \qquad \begin{array}{r} \times\,\Box \\ 6\,\overline{)\,12} \end{array}$$

$$\begin{array}{r} 2\,\overline{)\,12} \\ \times\,\Box \end{array} \qquad\qquad \begin{array}{r} 6\,\overline{)\,12} \\ \times\,\Box \end{array}$$

2007 - 2017 © Frank Ho, Amanda Ho, All rights reserved.　　www.homathchess.com

Different ways of writing multiplication (Learning division while doing multiplications)

$$\frac{14}{\square \times} = 2 \qquad\qquad \begin{array}{c} 2 \\ \times \\ 7 \end{array} \qquad\qquad \frac{14}{\square \times} = 7$$

$$2 \times \square \quad = \quad \boxed{} \quad = \quad 7 \times \square$$

$$
\begin{array}{c} \times \square \\ 2\,\overline{)\,14} \end{array}
\qquad\qquad
\begin{array}{c} 7 \\ \times \\ \square \end{array}
\qquad\qquad
\begin{array}{c} \times \square \\ 7\,\overline{)\,14} \end{array}
$$

$$
\begin{array}{c} 2\,\underline{)\,14} \\ \times \square \end{array}
\qquad\qquad\qquad\qquad
\begin{array}{c} 7\,\underline{)\,14} \\ \times \square \end{array}
$$

Ho Math Chess　何数棋谜　妈！我会棋谜式乘法啦！
Mom! I Learn Multiplication Using Math-Chess-Puzzles Connection!
Student's Name _____ Date _____
2007 - 2017 © Frank Ho, Amanda Ho, All rights reserved.　www.homathchess.com

Different ways of writing multiplication (Learning division while doing multiplications)

Different ways of writing multiplication (Learning division while doing multiplications)

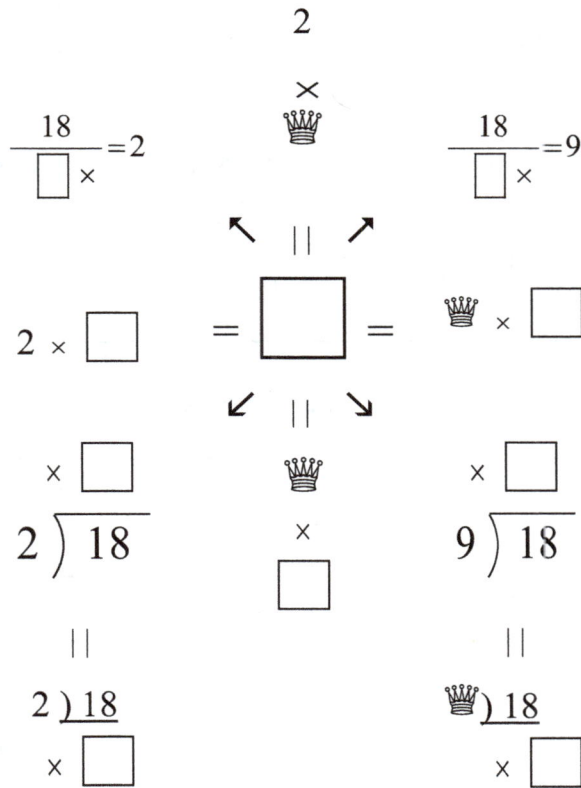

2 times

Fill in each □ with >, <, or =.

$8 \times 2 \;\square\; > \; 7 \times 2$	$8 \times 2 \;\square\; < \; 9 \times 2$	$8 \times 2 \;\square\; > \; 7 \times 2$
$1 \times 2 \;\square\; = \; 2 \times 1$	$4 \times 2 \;\square\; < \; 2 \times 5$	$2 \times 2 \;\square\; < \; 2 \times 3$
$2 \times 7 \;\square\; = \; 7 \times 2$	$5 \times 2 \;\square\; < \; 2 \times 6$	$5 \times 2 \;\square\; > \; 4 \times 2$
$8 \times 2 \;\square\; > \; 7 \times 2$	$9 \times 2 \;\square\; = \; 9 \times 2$	$3 \times 2 \;\square\; > \; 2 \times 2$
$8 \times 2 \;\square\; < \; 9 \times 2$	$5 \times 2 \;\square\; = \; 2 \times 5$	$6 \times 2 \;\square\; < \; 7 \times 2$
$8 \times 2 \;\square\; > \; 6 \times 2$	$6 \times 2 \;\square\; > \; 4 \times 2$	$6 \times 2 \;\square\; = \; 2 \times 6$
$8 \times 2 \;\square\; > \; 7 \times 2$	$9 \times 2 \;\square\; = \; 2 \times 9$	$2 \times 2 \;\square\; < \; 7 \times 2$
$3 \times 2 \;\square\; = \; 2 \times 3$	$8 \times 2 \;\square\; > \; 2 \times 5$	$2 \times 2 \;\square\; < \; 2 \times 3$
$9 \times 2 \;\square\; > \; 7 \times 2$	$4 \times 2 \;\square\; > \; 2 \times 3$	$8 \times 2 \;\square\; > \; 2 \times 4$
$8 \times 2 \;\square\; > \; 7 \times 2$	$5 \times 2 \;\square\; < \; 7 \times 2$	$6 \times 2 \;\square\; > \; 5 \times 2$

2007 - 2017 © Frank Ho, Amanda Ho, All rights reserved.   www.homathchess.com

Counting 3's multiples (Tripling)

Circle the following 3's multiples.

1 2 3 4 5 6 7 8 9 10 11 12 13 14 15 16 17 18
19 20 21 22 23 24 25 26 27

Fill in the following ☐ with a number.

Sequence	1	2	3	4	5	6	7	8	9
Add 3	☐	6	9	☐	15	18	☐	☐	27

Sequence	1	2	♗	4	♖	6	7	8	♕
Add 3	3	6	☐	☐	15	18	☐	☐	27

Sequence	1	2	♗	4	♖	6	7	8	♕
Add 3	☐	6	♕	☐	15	☐	☐	☐	27

Sequence	1	2	3	4	5	6	7	8	9
Add 3	☐	☐	9	☐	15	18	☐	☐	27

Sequence	1	2	♗	4	♖	6	7	8	♕
Add 3	☐	6	☐	☐	☐	18	☐	☐	27

2007 - 2017 © Frank Ho, Amanda Ho, All rights reserved.   www.homathchess.com

3 times

$2+2+2 =$ ☐	$= 3 \times 2 =$	$2 \times 3 =$ ☐	2 2 $+\ 2$ ☐ $= 3 \times 2 =$ ☐	
$3+3+3 =$ ☐	$= 3 \times 3 =$	♗ $\times$ ♗ $=$ ☐	♗ ♗ $+\ ♗$ ☐ $= 3 \times 3 =$ ☐	
$4+4+4 =$ ☐	$= 3 \times 4 =$	$4 \times 3 =$ ☐	4 4 $+\ 4$ ☐ $= 3 \times 4 =$ ☐	
$5+5+5 =$ ☐	$= 3 \times$ ♖ $=$	$5 \times 3 =$ ☐	♖ ♖ $+\ ♖$ ☐ $= 3 \times$ ♖ $=$ ☐	

2007 - 2017 © Frank Ho, Amanda Ho, All rights reserved.   www.homathchess.com

3 times

6+6+6 = $\square$	= 3 × 6 =	6 × 3 = $\square$	6 6 + 6 $\square$ = ♗ × 6 = $\square$
7+7+7 = $\square$	= 3 × 7 =	7 × 3 = $\square$	7 7 + 7 $\square$ = ♗ × 7 = $\square$
8+8+8 = $\square$	= 3 × 8 =	8 × 3 = $\square$	8 8 + 8 $\square$ = 3 × 8 = $\square$
9+9+9 = $\square$	= 3 × 9 =	9 × 3 = $\square$	9 9 + 9 $\square$ = 3 × ♕ = $\square$

2007 - 2017 © Frank Ho, Amanda Ho, All rights reserved.　　www.homathchess.com

3 times

3 × 1 = ☐	Three times one is ☐	1 × 3 = ☐	One times three is ☐
3 × 2 = ☐	Three times two is ☐	2 × 3 = ☐	Two times three is ☐
3 × 3 = ☐	Three times three is ☐	3 × 3 = ☐	Three times three is ☐
3 × 4 = ☐	Three times four is ☐	4 × 3 = ☐	Four times three is ☐
3 × 5 = ☐	Three times five is ☐	5 × 3 = ☐	Five times three is ☐
3 × 6 = ☐	Three times six is ☐	6 × 3 = ☐	Six times three is ☐
3 × 7 = ☐	**Three times seven is** ☐	7 × 3 = ☐	Seven times three is ☐
3 × 8 = ☐	Three times eight is ☐	8 × 3 = ☐	Eight times three is ☐
3 × 9 = ☐	Three times nine is ☐	♛ × 3 = ☐	Nine times three is ☐

```
  3        1        2        3        3
× 1      × 3      × 3      × 2      × 3
 ☐        ☐        ☐        ☐        ☐

  5        3        7        3        ♛
× 3      × 6      × 3      × 8      × 3
☐☐       ☐☐       ☐☐       ☐☐       ☐☐

  4        3        5        3        3
× 3      × 8      × 3      × 6      × 9
☐☐       ☐☐       ☐☐       ☐☐       ☐☐
```

42

Ho Math Chess 何数棋谜 妈！我会棋谜式乘法啦！

Mom! I Learn Multiplication Using Math-Chess-Puzzles Connection!

Student's Name _____ Date _____

2007 - 2017 © Frank Ho, Amanda Ho, All rights reserved.   www.homathchess.com

3   1   4   3   3
X 1   X 3   X ♗   X 4   X 3

♖   3   ♘   8   ♗
X 3   X 6   X 7   X 3   X 9

4   3   5   6   7
X ♘   X 4   X 3   X ♗   X ♘

8   ♕   3   ♘   3
X 3   X 3   X 4   X 8   X 5

4   3   3   3   3
X 3   X 6   X 7   X 3   X 9

2007 - 2017 © Frank Ho, Amanda Ho, All rights reserved.   www.homathchess.com

Oral practice

Three one three	3 1 ☐	$\begin{array}{r} 1\,1 \\ \times\quad 3 \\ \hline 33\ \square\ \square \end{array}$
Three two six (is half-dozen)	3 2 ☐	$\begin{array}{r} 2\,2 \\ \times\quad 3 \\ \hline 66\ \square\ \square \end{array}$
Three three nine	3 3 ☐	$\begin{array}{r} 3\,3 \\ \times\quad ♘ \\ \hline 99\ \square\ \square \end{array}$
Three four twelve (is a dozen)	3 4 ☐	$\begin{array}{r} {}^{1}\ \ \\ 4\,4 \\ \times\quad 3 \\ \hline 132\ \square\ \square\ \square \end{array}$
Three five fifteen	3 5 ☐	$\begin{array}{r} {}^{1}\ \ \\ 5\,5 \\ \times\quad ♘ \\ \hline 165\ \square\ \square\ \square \end{array}$

2007 - 2017 © Frank Ho, Amanda Ho, All rights reserved.　　www.homathchess.com

Oral practice

Three six eighteen	♗ 6 ☐	$$\begin{array}{r} {}^{1} \\ 6\,6 \\ \times\ \ 3 \\ \hline 198\ \square\square\square \end{array}$$
Three seven twenty-one	3 7 ☐	$$\begin{array}{r} 7\,7 \\ \times\ \ ♗ \\ \hline 231\ \square\square\square \end{array}$$
Three eight twenty-four	3 8 ☐	$$\begin{array}{r} 8\,8 \\ \times\ \ 3 \\ \hline 264\ \square\square\square \end{array}$$
Three nine twenty-seven	♘ 9 ☐	$$\begin{array}{r} 9\,9 \\ \times\ \ 3 \\ \hline 297\ \square\square\square \end{array}$$
Three five fifteen	3 5 ☐	$$\begin{array}{r} 5\,5 \\ \times\ \ 3 \\ \hline 165\ \square\square\square \end{array}$$
Three six eighteen	3 6 ☐	$$\begin{array}{r} 6\,6 \\ \times\ \ ♘ \\ \hline 198\ \square\square\square \end{array}$$

2007 - 2017 © Frank Ho, Amanda Ho, All rights reserved.   www.homathchess.com

Preparing for division

☐	☐	☐	☐	☐
X 2	X ♘	X 4	X 2	X ♗
6	♕	1 2	6	1 2

☐	☐	☐	☐	☐
X 4	X 6	X 3	X 3	X 2
2 8	1 8	2 1	1 5	6

☐	☐	☐	☐	☐
X 9	X 3	X ♗	X ♘	X 7
2 7	1 2	2 1	1 5	2 1

☐	☐	☐	☐	☐
X ♖	X 3	X 4	X ♘	X 2
1 5	6	1 2	♕	6

☐	☐	☐	☐	☐
X 4	X 5	X ♗	X 2	X 3
1 2	1 5	1 8	1 2	2 1

2007 - 2017 © Frank Ho, Amanda Ho, All rights reserved.   www.homathchess.com

Preparing for division

☐ X ♞ = 3	X ☐ 3)3	☐)3____ X ♞
☐ X 3 = 6	X ☐ 3)6	☐)6____ X 3
☐ X ♗ = 9	X ☐ 3)9	☐)♛____ X ♛
☐ X 3 = 12	X ☐ 3)12	☐)12____ X 3
☐ X ♞ = 15	X ☐ 3)15	☐)15____ X ♗
☐ X ♗ = 18	X ☐ 3)18	☐)18____ X ♞
☐ X ♞ = 21	X ☐ 3)21	☐)21____ X ♞

Fill in ☐ with answer.

Times	Grouping	Addition
3 × $1 = ☐	3 of ☐ = ♗	$1 + $ 1 + $ 1 = ☐
1 × $3 = ☐	1 of ☐ = 3	$3 = ☐

Fill in ☐ with answer.

Expression	Grouping	Addition
3 × $2	3 of ☐ = 6	$2 + $2 + $2 = ☐
2 × $3	2 of ☐ = 6	$3 + $3 = ☐

3 × 1 = ☐ = 1 × 3 = ☐	1 × 3 = ☐ = 3 × 1 = ☐
3 × ☐ = 6 = 2 × ☐ = ☐	2 × ☐ = 6 = 3 × ☐ = ☐

3 1 ☐	3 5 ☐	3 9 ☐	♗ 4 ☐	3 8 ☐
3 2 ☐	♗ 6 ☐	♗ 1 ☐	♗ ♖ ☐	♘ 9 ☐
♘ ♗ ☐	3 7 ☐	3 2 ☐	3 6 ☐	3 1 ☐
3 4 ☐	3 8 ☐	3 ♘ ☐	3 7 ☐	3 2 ☐

Student's Name _____ Date _____

2007 - 2017 © Frank Ho, Amanda Ho, All rights reserved.　　www.homathchess.com

Fill in ☐ with answer.

Times	Grouping	Addition
$3 \times \$3 = \square$	3 of $\square$ = 9	$\$3 + \$3 + \$3 = \square$
$3 \times \$3 = \square$	3 of $\square$ = 9	$\$3 + \$3 + \$♗ = \square$

Fill in ☐ with answer.

Expression	Grouping	Addition
$3 \times \$4$	♗ of $\square$ = 12	$\$4 + \$4 + \$4 = \square$
$4 \times \$3$	4 of $\square$ = 12	$\$3 + \$3 + \$3 + \$♘ = \square$

♘ $\times 3 = \square = 3 \times 3 = \square$	♗ $\times 3 = \square = 3 \times 3 = \square$
$3 \times \square = 12 = 4 \times \square = \square$	$4 \times \square = 12 = 3 \times \square = \square$

3 1 ☐	3 5 ☐	♗ 9 ☐	3 4 ☐	3 8 ☐
3 2 ☐	3 6 ☐	3 1 ☐	♘ 5 ☐	3 9 ☐
3 3 ☐	♘ 7 ☐	3 2 ☐	3 6 ☐	3 1 ☐
3 4 ☐	3 8 ☐	♗ ♗ ☐	3 7 ☐	♘ 2 ☐

Ho Math Chess 何数棋谜 妈！我会棋谜式乘法啦！
Mom! I Learn Multiplication Using Math-Chess-Puzzles Connection!

Student's Name _____ Date _____

2007 - 2017 © Frank Ho, Amanda Ho, All rights reserved.   www.homathchess.com

Fill in ☐ with answer.

Times	Grouping	Addition
3 × $5 = ☐	3 of ☐ = 15	$5 + $♖ + $5 = ☐
5 × $♘ = ☐	5 of ☐ = 15	$3 + $3 + $♗ + $3 + $3 = ☐

Times	Grouping	Addition
3 × $6 = ☐	3 of ☐ = 18	$6 + $6 + $6 = ☐
6 × $3 = ☐	6 of ☐ = 18	$3 + $3 + $3 + $3 + $3 + $3 = ☐

♘ × 5 = ☐ = 5 × 3 = ☐	♖ × 3 = ☐ = ♘ × 5 = ☐
3 × ☐ = 18 = 6 × ☐ = ☐	6 × ☐ = 18 = ♗ × ☐ = ☐

♘ ♙ ☐	3 5 ☐	3 9 ☐	3 4 ☐	♗ 8 ☐
3 2 ☐	3 6 ☐	♘ 1 ☐	3 ♖ ☐	3 9 ☐
3 3 ☐	♘ 7 ☐	3 2 ☐	3 6 ☐	♘ 1 ☐
♗ 4 ☐	3 8 ☐	3 3 ☐	♘ 7 ☐	3 2 ☐

Ho Math Chess　何数棋谜　妈！我会棋谜式乘法啦！

Mom! I Learn Multiplication Using Math-Chess-Puzzles Connection!

Student's Name _____ Date _____

2007 - 2017 © Frank Ho, Amanda Ho, All rights reserved.　　www.homathchess.com

Fill in ☐ with answer.

Times	Grouping	Addition
$3 \times \$7 = \square$	$3 \text{ of } \square = 21$	$\$7 + \$7 + \$7 = \square$
$7 \times \$3 = \square$	$7 \text{ of } \square = 21$	$\$3 + \$3 + \$3 + \$3 + \$3 + \$3 + \$3 = \square$

Times	Grouping	Addition
$3 \times \$8 = \square$	$3 \text{ of } \square = 24$	$\$8 + \$8 + \$8 = \square$
$8 \times \$3 = \square$	$8 \text{ of } \square = 24$	$\$3 + \$3 + \$3 + \$3 + \$3 + \$3 + \$3 + \$3 = \square$

$3 \times 7 = \square = 7 \times ♟ = \square$	$7 \times 3 = \square = 3 \times 7 = \square$
$3 \times \square = 24 = 8 \times \square = \square$	$8 \times \square = 24 = ♟ \times \square = \square$

3 1 ☐	3 5 ☐	3 9 ☐	3 4 ☐	3 8 ☐
3 2 ☐	♟ 6 ☐	3 ♙ ☐	♟ 5 ☐	3 9 ☐
3 ♟ ☐	3 7 ☐	3 2 ☐	3 6 ☐	3 ♙ ☐
3 4 ☐	3 8 ☐	♟ ♟ ☐	3 7 ☐	3 2 ☐

2007 - 2017 © Frank Ho, Amanda Ho, All rights reserved.　www.homathchess.com

Fill in ☐ with answer.

Times	Grouping	Addition
$3 \times \$9 = \square$	3 of $\square = 27$	$\$9 + \$9 + \$9 = \square$
$9 \times \$3 = \square$	9 of $\square = 27$	$\$3 + \$3 + \$3 + \$3 + \$3 + \$3 + \$3 + \$3 + \$3 = \square$

Times	Grouping	Addition
$3 \times \$8 = \square$	3 of $\square = 24$	$\$8 + \$8 + \$8 = \square$
$8 \times \$3 = \square$	8 of $\square = 24$	$\$3 + \$3 + \$3 + \$3 + \$3 + \$3 + \$3 + \$3 = \square$

$3 \times 8 = \square = 8 \times 3 = \square$	$8 \times 3 = \square = 3 \times 8 = \square$
$3 \times \square = 27 = 9 \times \square = \square$	♛ $\times \square = 27 = 3 \times \square = \square$

3 1 ☐	3 5 ☐	3 ♛ ☐	3 4 ☐	♞ 8 ☐
3 2 ☐	♗ 6 ☐	3 1 ☐	♗ 5 ☐	3 9 ☐
3 ♗ ☐	3 7 ☐	♗ 2 ☐	3 6 ☐	3 ♙ ☐
3 4 ☐	3 8 ☐	3 3 ☐	♗ 7 ☐	3 2 ☐

2007 - 2017 © Frank Ho, Amanda Ho, All rights reserved.　　www.homathchess.com

Preparing for division

☐ X ♗ = 24	X ☐ 3)24	☐)24 X 3
☐ X 3 = 27	X ☐ 3)27	☐)27 X 3
☐ X ♗ = ♛	X ☐ 3)9	☐)♛ X 3
☐ X 3 = 12	X ☐ 3)12	☐)12 X 3
☐ X ♗ = 15	X ☐ 3)15	☐)15 X ♞
☐ X 3 = 18	X ☐ 3)18	☐)18 X 3
☐ X 3 = 21	X ☐ 3)21	☐)21 X ♞

2007 - 2017 © Frank Ho, Amanda Ho, All rights reserved.   www.homathchess.com

Cross multiplication

12 ↖ ↗ 12 $\dfrac{6}{2} = \dfrac{6}{2}$	□ 6 ↖ ↗ $\dfrac{3}{3} = \dfrac{2}{2}$	□ 9 ↖ ↗ $\dfrac{3}{3} = \dfrac{3}{3}$	□ 12 ↖ ↗ $\dfrac{3}{3} = \dfrac{4}{4}$
□ 15 ↖ ↗ $\dfrac{3}{3} = \dfrac{5}{5}$	□ 18 ↖ ↗ $\dfrac{3}{3} = \dfrac{6}{6}$	□ 21 ↖ ↗ $\dfrac{3}{3} = \dfrac{7}{7}$	□ 18 ↖ ↗ $\dfrac{3}{3} = \dfrac{9}{9}$
□ 15 ↖ ↗ $\dfrac{3}{3} = \dfrac{5}{5}$	□ 12 ↖ ↗ $\dfrac{3}{3} = \dfrac{4}{4}$	□ 21 ↖ ↗ $\dfrac{3}{3} = \dfrac{7}{7}$	□ 24 ↖ ↗ $\dfrac{3}{3} = \dfrac{8}{8}$
□ 18 ↖ ↗ $\dfrac{3}{3} = \dfrac{6}{6}$	□ 24 ↖ ↗ $\dfrac{3}{3} = \dfrac{8}{8}$	□ 27 ↖ ↗ $\dfrac{3}{3} = \dfrac{9}{9}$	□ 9 ↖ ↗ $\dfrac{3}{3} = \dfrac{3}{3}$

2007 - 2017 © Frank Ho, Amanda Ho, All rights reserved.　　www.homathchess.com

Different ways of writing multiplication (Learning division while doing multiplications)

$$\frac{6}{\square \times} = 2 \qquad \overset{\times}{2} \qquad \frac{6}{\square \times} = 3$$

$$2 \times \square \;=\; \boxed{} \;=\; \text{♝} \times \square$$

$$2\,\overline{)\,6} \qquad \overset{2}{\underset{\square}{\times}} \qquad 3\,\overline{)\,6}$$

$$2\,\overline{)\,6} \qquad\qquad ♞\,\overline{)\,6}$$
$$\times \square \qquad\qquad \times \square$$

2007 - 2017 © Frank Ho, Amanda Ho, All rights reserved.　　www.homathchess.com

Different ways of writing multiplication (Learning division while doing multiplications)

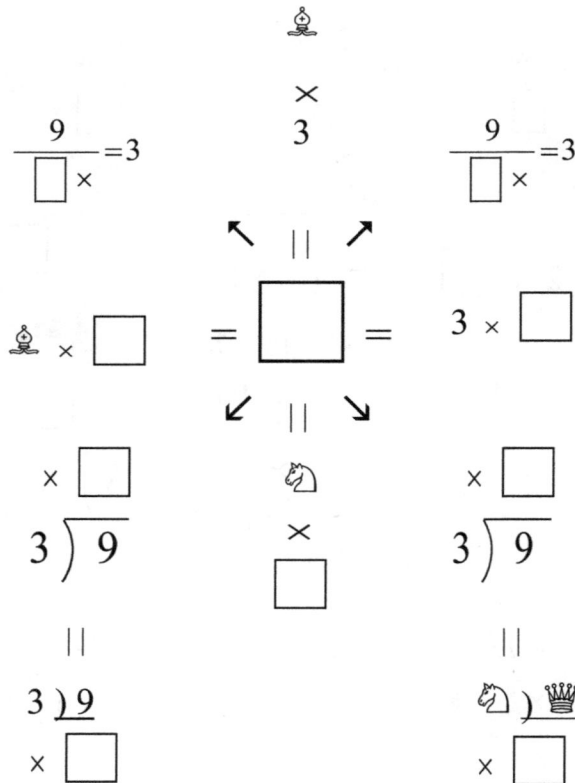

$$\frac{9}{\square \times} = 3$$

$$♗ \times 3$$

$$\frac{9}{\square \times} = 3$$

$$♗ \times \square = \boxed{} = 3 \times \square$$

$$\times \square$$
$$3\,\overline{)\,9}$$

$$♘ \times \square$$

$$\times \square$$
$$3\,\overline{)\,9}$$

$$3\,\overline{)\,9}$$
$$\times \square$$

$$♘\,\overline{)\,♕}$$
$$\times \square$$

2007 - 2017 © Frank Ho, Amanda Ho, All rights reserved.   www.homathchess.com

Different ways of writing multiplication (Learning division while doing multiplications)

$$\frac{12}{\boxed{}\times}=3 \qquad \overset{\times}{4} \qquad \frac{12}{\boxed{}\times}=4$$

$$\nwarrow \;\; || \;\; \nearrow$$

$$3 \times \boxed{} \quad = \quad \boxed{} \quad = \quad \boxed{} \times 4$$

$$\swarrow \;\; || \;\; \searrow$$

$$\times \boxed{} \qquad\qquad 4 \qquad\qquad \times \boxed{}$$

$$3\,\overline{)\,12} \qquad \overset{\times}{\underset{\boxed{}}{}} \qquad 4\,\overline{)\,12}$$

$$|| \qquad\qquad\qquad\qquad ||$$

$$♘\,\overline{)\,12} \qquad\qquad\qquad 4\,\overline{)\,12}$$

$$\times \boxed{} \qquad\qquad\qquad \times \boxed{}$$

2007 - 2017 © Frank Ho, Amanda Ho, All rights reserved.   www.homathchess.com

Different ways of writing multiplication (Learning division while doing multiplications)

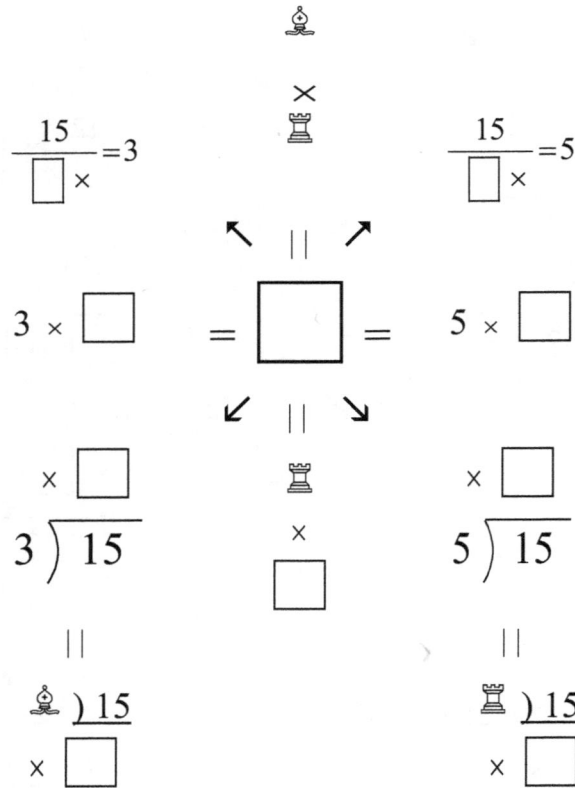

2007 - 2017 © Frank Ho, Amanda Ho, All rights reserved.   www.homathchess.com

Different ways of writing multiplication (Learning division while doing multiplications)

$$\frac{18}{\Box\times}=3 \qquad \times \qquad \frac{18}{\Box\times}=6$$

$$6$$

$$3\times\Box = \Box = 6\times\Box$$

$$\times\Box \qquad 6 \qquad \times\Box$$

$$3\overline{)18} \qquad \times \qquad 6\overline{)18}$$

$$\Box$$

$$\text{♞}\overline{)18} \qquad 6\overline{)18}$$
$$\times\Box \qquad \times\Box$$

2007 - 2017 © Frank Ho, Amanda Ho, All rights reserved.   www.homathchess.com

Different ways of writing multiplication (Learning division while doing multiplications)

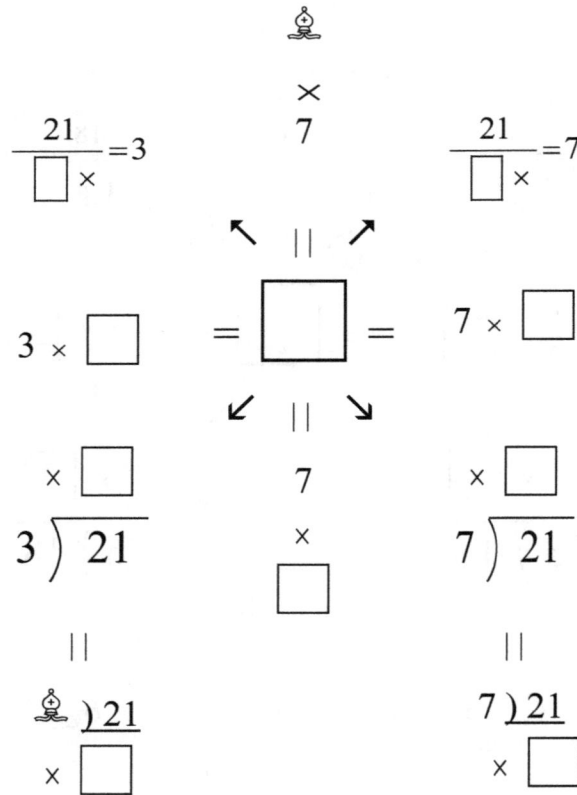

$$\frac{21}{\square\times}=3 \qquad \times \atop 7 \qquad \frac{21}{\square\times}=7$$

$$3 \times \square \quad = \boxed{} = \quad 7 \times \square$$

$$\times\,\square \qquad 7 \qquad \times\,\square$$

$$3\overline{)\,21} \qquad \times \atop \square \qquad 7\overline{)\,21}$$

$$||\qquad\qquad\qquad||$$

$$\overset{\text{♗}}{}\overline{)\,21} \qquad\qquad 7\,\overline{)\,21}$$

$$\times\,\square \qquad\qquad\qquad \times\,\square$$

2007 - 2017 © Frank Ho, Amanda Ho, All rights reserved.　www.homathchess.com

Different ways of writing multiplication (Learning division while doing multiplications)

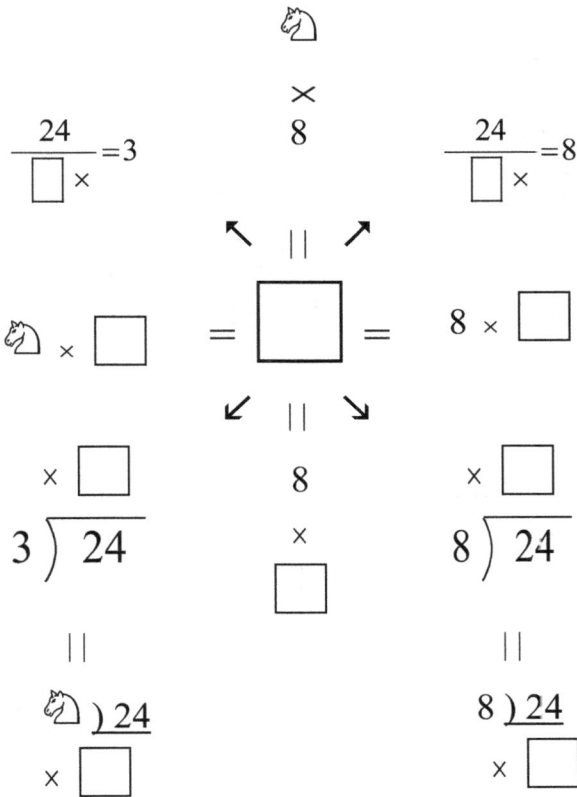

Different ways of writing multiplication (Learning division while doing multiplications)

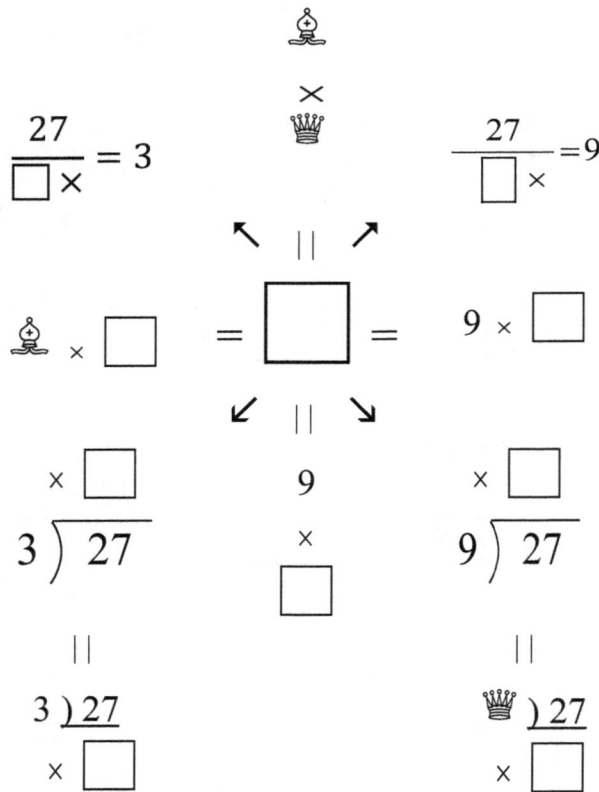

Ho Math Chess 何数棋谜 妈！我会棋谜式乘法啦！
Mom! I Learn Multiplication Using Math-Chess-Puzzles Connection!

Student's Name _____ Date _____

2007 - 2017 © Frank Ho, Amanda Ho, All rights reserved. www.homathchess.com

3 times

Fill in each ☐ with >, <, or =.

8 × 3 ☐ > 7 × 3	6 × 3 ☐ > 5 × 3	8 × 3 ☐ < 9 × 3
1 × 3 ☐ = ♞ × 1	9 × 3 ☐ = 3 × ♛	4 × 3 ☐ < 3 × 5
3 × 7 ☐ = 7 × ♝	5 × 3 ☐ > 4 × ♞	♜ × 3 ☐ < 3 × 6
8 × 3 ☐ > 7 × 3	8 × 3 ☐ > 7 × ♝	9 × 3 ☐ = 9 × 3
8 × ♝ ☐ < 9 × 3	3 × 2 ☐ = 2 × 3	5 × 3 ☐ = ♞ × ♜
8 × 3 ☐ > 6 × 3	6 × ♝ ☐ = 3 × 6	6 × 3 ☐ > 4 × 3
8 × 3 ☐ > 7 × ♝	2 × 3 ☐ < 7 × ♞	9 × ♞ ☐ = 3 × 9
3 × 3 ☐ > 2 × 3	3 × 3 ☐ = 3 × 3	8 × 3 ☐ > 3 × ♜
9 × ♞ ☐ > 7 × 3	8 × 3 ☐ > 3 × 4	4 × ♝ ☐ > 3 × 3
8 × 3 ☐ > 7 × 3	6 × 3 ☐ > ♜ × ♞	5 × 3 ☐ < 7 × ♝

2007 - 2017 © Frank Ho, Amanda Ho, All rights reserved.   www.homathchess.com

Counting 4's multiples

Write 4's multiples.

4, 8, 12, ☐ , ☐ , ☐ , ☐ , ☐ , ☐

Fill in the following each ☐ with a number.

Sequence	1	2	3	4	♖	6	7	8	♕
Add 4	☐	8	☐	16	☐	24	☐	32	☐

Sequence	♙	2	♗	4	5	6	7	8	9
Add 4	4	☐	12	☐	20	☐	28	☐	36

Sequence	1	2	3	4	♖	6	7	8	9
Add 4	☐	8	☐	16	☐	24	☐	32	☐

Sequence	♙	2	♗	4	5	6	7	8	♕
Add 4	4	☐	12	☐	20	☐	28	☐	36

Sequence	1	2	3	4	♖	6	7	8	9
Add 4	☐	☐	☐	16	☐	24	☐	32	☐

Sequence	♙	2	♗	4	5	6	7	8	9
Add 4	☐	8	☐	☐	☐	24	☐	32	☐

2007 - 2017 © Frank Ho, Amanda Ho, All rights reserved.　　www.homathchess.com

4 times

$4 \times 1 = \square$	Four times one is $\square$	$1 \times 3 = \square$	One times four is $\square$
$4 \times 2 = \square$	Four times two is $\square$	$2 \times 3 = \square$	Two times four is $\square$
$4 \times 3 = \square$	Four times three is $\square$	♟ $\times 3 = \square$	Three times four is $\square$
$4 \times 4 = \square$	Four times four is $\square$	$4 \times 3 = \square$	Four times four is $\square$
$4 \times ♜ = \square$	Four times five is $\square$	♜ $\times 3 = \square$	Five times four is $\square$
$4 \times 6 = \square$	Four times six is $\square$	$6 \times 3 = \square$	Six times four is $\square$
$4 \times 7 = \square$	Four times seven is $\square$	$7 \times 3 = \square$	Seven times four is $\square$
$4 \times 8 = \square$	Four times eight is $\square$	$8 \times 3 = \square$	Eight times four is $\square$
$4 \times 9 = \square$	Four times nine is $\square$	$9 \times ♟ = \square$	Nine times four is $\square$

4	♙	2	4	4
X 1	X 4	X 4	X 2	X ♝
$\square$	$\square$	$\square$	$\square$	$\square$

4	4	7	4	♛
X 3	X 6	X 4	X 8	X 4
☐☐	☐☐	☐☐	☐☐	☐☐

4	4	♜	4	4
X ♜	X 8	X 4	X 6	X ♛
☐☐	☐☐	☐☐	☐☐	☐☐

Mom! I Learn Multiplication Using Math-Chess-Puzzles Connection!

Student's Name _____ Date _____

2007 - 2017 © Frank Ho, Amanda Ho, All rights reserved.   www.homathchess.com

4 × ♙	1 × 4	4 × 2	2 × 4	4 × ♗
5 × 4	4 × 6	4 × 7	8 × 4	4 × 9
4 × 6	♖ × 4	8 × 4	4 × 6	♕ × 4
5 × 4	4 × 6	7 × 4	4 × 8	9 × 4
4 × ♖	4 × 8	4 × 2	4 × 6	4 × ♕

2007 - 2017 © Frank Ho, Amanda Ho, All rights reserved.   www.homathchess.com

Oral practice

four one four	4 1 ☐	$\begin{array}{r} 1\ 1 \\ \times\quad 4 \\ \hline 44\ \square\ \square \end{array}$
four two eight	4 2 ☐	$\begin{array}{r} 2\ 2 \\ \times\quad 4 \\ \hline 88\ \square\ \square \end{array}$
four three twelve (is a dozen)	4 3 ☐	$\begin{array}{r} 3\ 3 \\ \times\quad 4 \\ \hline 132\ \square\ \square\ \square \end{array}$
four four sixteen	4 4 ☐	$\begin{array}{r} {}^{1}\ \\ 4\ 4 \\ \times\quad 4 \\ \hline 176\ \square\ \square\ \square \end{array}$
four five twenty	4 ♖ ☐	$\begin{array}{r} {}^{2}\ \\ 5\ 5 \\ \times\quad 4 \\ \hline 220\ \square\ \square\ \square \end{array}$

Oral practice

four six twenty-four	4 6 ☐	$\begin{array}{r} ^2 \\ 6\,6 \\ \times\ \ \ 4 \\ \hline 264\,\square\square\square \end{array}$
four seven twenty-eight	4 7 ☐	$\begin{array}{r} 7\,7 \\ \times\ \ \ 4 \\ \hline 308\,\square\square\square \end{array}$
four eight thirty-two	4 8 ☐	$\begin{array}{r} 8\,8 \\ \times\ \ \ 4 \\ \hline 352\,\square\square\square \end{array}$
four nine thirty-six	4 ♛ ☐	$\begin{array}{r} 9\,9 \\ \times\ \ \ 4 \\ \hline 396\,\square\square\square \end{array}$
four five twenty	4 ♜ ☐	$\begin{array}{r} 5\,5 \\ \times\ \ \ 4 \\ \hline 220\,\square\square\square \end{array}$
four six twenty-four	4 6 ☐	$\begin{array}{r} 6\,6 \\ \times\ \ \ 4 \\ \hline 264\,\square\square\square \end{array}$

Ho Math Chess   何数棋谜   妈!我会棋谜式乘法啦!
Mom! I Learn Multiplication Using Math-Chess-Puzzles Connection!

Student's Name _____ Date _____

2007 - 2017 © Frank Ho, Amanda Ho, All rights reserved.   www.homathchess.com

Fill in ☐ with answer.

Times	Grouping	Addition
$4 \times \$1 = \square$	$4 \text{ of } \square = 4$	$\$1 + \$1 + \$1 + \$1 = \square$
$1 \times \$4 = \square$	$1 \text{ of } \square = 4$	$\$4$

Fill in ☐ with answer.

Expression	Grouping	Addition
$4 \times \$2$	$4 \text{ of } \square = 8$	$\$2 + \$2 + \$2 + \$2 = \square$
$2 \times \$4$	$2 \text{ of } \square = 8$	$\$4 + \$4 = \square$

$4 \times 1 = \square = 1 \times 4 = \square$	$1 \times 4 = \square = 4 \times 1 = \square$
$4 \times \square = 8 = 2 \times \square = \square$	$2 \times \square = 8 = 4 \times \square = \square$

4 1 ☐	4 5 ☐	4 9 ☐	4 4 ☐	4 8 ☐
4 2 ☐	4 6 ☐	4 1 ☐	4 5 ☐	4 ♕ ☐
4 3 ☐	4 7 ☐	4 2 ☐	4 6 ☐	4 1 ☐
4 4 ☐	4 8 ☐	4 3 ☐	4 7 ☐	4 2 ☐

Fill in ☐ with answer.

Times	Grouping	Addition
$4 \times \$3 = $ ☐	4 of ☐ $= 12$	$\$3 + \$3 + \$3 + \$3 = $ ☐
$3 \times \$4 = $ ☐	3 of ☐ $= 12$	$\$4 + \$4 + \$4 = $ ☐

Fill in ☐ with answer.

Expression	Grouping	Addition
$4 \times \$4$	4 of ☐ $= 16$	$\$4 + \$4 + \$4 + \$4 = $ ☐
$4 \times \$4$	4 of ☐ $= 16$	$\$4 + \$4 + \$4 + \$4 = $ ☐

$3 \times 4 = $ ☐ $= 4 \times 3 = $ ☐	$4 \times 3 = $ ☐ $= 3 \times 4 = $ ☐
$4 \times $ ☐ $= 16 = 4 \times $ ☐ $= $ ☐	$4 \times $ ☐ $= 16 = 4 \times $ ☐ $= $ ☐

4 ♟ ☐	4 5 ☐	4 9 ☐	4 4 ☐	4 8 ☐
4 2 ☐	4 6 ☐	4 ♟ ☐	4 5 ☐	4 ♛ ☐
4 ♞ ☐	4 7 ☐	4 2 ☐	4 6 ☐	4 1 ☐
4 4 ☐	4 8 ☐	4 ♞ ☐	4 7 ☐	4 2 ☐

Student's Name _____ Date _____

2007 - 2017 © Frank Ho, Amanda Ho, All rights reserved.   www.homathchess.com

Fill in ☐ with answer.

Times	Grouping	Addition
$4 \times \$5 = \square$	4 of $\square$ = 20	$\$5 + \$5 + \$5 + \$5 = \square$
$5 \times \$4 = \square$	5 of $\square$ = 20	$\$4 + \$4 + \$4 + \$4 + \$4 = \square$

Times	Grouping	Addition
$4 \times \$6 = \square$	4 of $\square$ = 24	$\$6 + \$6 + \$6 + \$6 = \square$
$6 \times \$4 = \square$	6 of $\square$ = 24	$\$4 + \$4 + \$4 + \$4 + \$4 + \$4 = \square$

$4 \times 5 = \square = ♖ \times 4 = \square$	$♖ \times 4 = \square = 4 \times 5 = \square$
$4 \times \square = 24 = 6 \times \square = \square$	$6 \times \square = 24 = 6 \times \square = \square$

4 1 ☐	4 5 ☐	4 9 ☐	4 4 ☐	4 8 ☐
4 2 ☐	4 6 ☐	4 1 ☐	4 ♖ ☐	4 9 ☐
4 ♗ ☐	4 7 ☐	4 2 ☐	4 6 ☐	4 ♙ ☐
4 4 ☐	4 8 ☐	4 3 ☐	4 7 ☐	4 2 ☐

Ho Math Chess   何数棋谜   妈！我会棋谜式乘法啦！
Mom! I Learn Multiplication Using Math-Chess-Puzzles Connection!

Student's Name _____ Date _____

2007 - 2017 © Frank Ho, Amanda Ho, All rights reserved.   www.homathchess.com

Fill in _____ and ☐ with answers.

Times	Grouping	Addition
$4 \times \$7 = $ ☐	4 of ☐ = 28	$\$7 + \$7 + \$7 + \$7 = $ ☐
$7 \times \$4 = $ ☐	7 of ☐ = 28	$\$4 + \$4 + \$4 + \$4 + \$4 + \$4 + \$4 = $ ☐

Times	Grouping	Addition
$4 \times \$8 = $ ☐	4 of ☐ = 32	$\$8 + \$8 + \$8 + \$8 = $ ☐
$8 \times \$4 = $ ☐	8 of ☐ = 32	$\$4 + \$4 + \$4 + \$4 + \$4 + \$4 + \$4 + \$4 = $ ☐

$4 \times 7 = $ ☐ $= 7 \times 4 = $ ☐	$7 \times 4 = $ ☐ $= 4 \times 7 = $ ☐
$4 \times $ ☐ $= 32 = 8 \times $ ☐ $= $ ☐	$8 \times $ ☐ $= 32 = 4 \times $ ☐ $= $ ☐

4 1 ☐	4 5 ☐	4 9 ☐	4 4 ☐	4 8 ☐
4 2 ☐	4 6 ☐	4 1 ☐	4 ♖ ☐	4 9 ☐
4 3 ☐	4 7 ☐	4 2 ☐	4 6 ☐	4 1 ☐
4 4 ☐	4 8 ☐	4 ♗ ☐	4 7 ☐	4 2 ☐

72

2007 - 2017 © Frank Ho, Amanda Ho, All rights reserved.　www.homathchess.com

Fill in _____ and ☐ with answers.

Times	Grouping	Addition
$4 \times \$9 = \square$	4 of $\square = 36$	$\$9 + \$9 + \$9 + \$9 = \square$
$9 \times \$4 = \square$	9 of $\square = 36$	$\$4 + \$4 + \$4 + \$4 + \$4 + \$4 + \$4 + \$4 + \$4 = \square$

Times	Grouping	Addition
$4 \times \$8 = \square$	4 of $\square = 32$	$\$8 + \$8 + \$8 + \$8 = \square$
$8 \times \$4 = \square$	8 of $\square = 32$	$\$4 + \$4 + \$4 + \$4 + \$4 + \$4 + \$4 + \$4 = \square$

$4 \times 9 = \square = ♕ \times 4 = \square$	$9 \times 4 = \square = 4 \times ♕ = \square$
$4 \times \square = 36 = 9 \times \square = \square$	$9 \times \square = 36 = 4 \times \square = \square$

$4\ 1\ \square$	$4\ 5\ \square$	$4\ ♕\ \square$	$4\ 4\ \square$	$4\ 8\ \square$
$4\ 2\ \square$	$4\ 6\ \square$	$4\ 1\ \square$	$4\ 5\ \square$	$4\ ♕\ \square$
$4\ ♗\ \square$	$4\ 7\ \square$	$4\ 2\ \square$	$4\ 6\ \square$	$4\ 1\ \square$
$4\ 4\ \square$	$4\ 8\ \square$	$4\ ♘\ \square$	$4\ 7\ \square$	$4\ 2\ \square$

Preparing for division

□	□	□	□	□
X 2	X 3	X 4	X 2	X 4
8	1 2	1 6	8	2 4

□	□	□	□	□
X 9	X 8	X 4	X ♞	X 4
3 6	3 2	2 8	1 2	1 6

□	□	□	□	□
X ♛	X 4	X 4	X 4	X 7
3 6	2 0	2 8	2 0	2 8

□	□	□	□	□
X 5	X ♞	X 4	X 4	X 2
2 0	1 2	3 2	3 6	8

□	□	□	□	□
X 4	X ♜	X 6	X 4	X 3
1 6	2 0	2 4	1 2	1 2

2007 - 2017 © Frank Ho, Amanda Ho, All rights reserved.   www.homathchess.com

Preparing for division

☐ X 4 = 4	X ☐ 4)4	☐)4 X 4
☐ X 4 = 8	X ☐ 4)8	☐)8 X 4
☐ X 4 = 12	X ☐ 3)12	☐)12 X 4
☐ X 4 = 16	X ☐ 4)16	☐)16 X 4
☐ X 4 = 20	X ☐ 4)20	☐)20 X 4
☐ X 4 = 24	X ☐ 4)24	☐)24 X 4
☐ X 4 = 28	X ☐ 4)28	☐)28 X 4

Preparing for division

☐ X 4 = 32	X ☐ 4⟌32	☐) 32 X 4
☐ X 4 = 36	X ☐ 4⟌36	☐) 36 X 4
☐ X 4 = 12	X ☐ 3⟌12	☐) 12 X 4
☐ X 4 = 16	X ☐ 4⟌16	☐) 16 X 4
☐ X 4 = 20	X ☐ 4⟌20	☐) 20 X 4
☐ X 4 = 24	X ☐ 4⟌24	☐) 24 X 4
☐ X 4 = 28	X ☐ 4⟌28	☐) 28 X 4

2007 - 2017 © Frank Ho, Amanda Ho, All rights reserved.　　www.homathchess.com

Cross multiplication

12　　12 ↖　↗ $\frac{6}{2} = \frac{6}{2}$	☐　**8** ↖　↗ $\frac{4}{4} = \frac{2}{2}$	☐　**12** ↖　↗ $\frac{4}{4} = \frac{3}{3}$	☐　**16** ↖　↗ $\frac{4}{4} = \frac{4}{4}$
☐　**20** ↖　↗ $\frac{4}{4} = \frac{5}{5}$	☐　**24** ↖　↗ $\frac{4}{4} = \frac{6}{6}$	☐　**28** ↖　↗ $\frac{4}{4} = \frac{7}{7}$	☐　**36** ↖　↗ $\frac{4}{4} = \frac{9}{9}$
☐　**20** ↖　↗ $\frac{4}{4} = \frac{5}{5}$	☐　**16** ↖　↗ $\frac{4}{4} = \frac{4}{4}$	☐　**28** ↖　↗ $\frac{4}{4} = \frac{7}{7}$	☐　**32** ↖　↗ $\frac{4}{4} = \frac{8}{8}$
☐　**24** ↖　↗ $\frac{4}{4} = \frac{6}{6}$	☐　**32** ↖　↗ $\frac{4}{4} = \frac{8}{8}$	☐　**36** ↖　↗ $\frac{4}{4} = \frac{9}{9}$	☐　**12** ↖　↗ $\frac{4}{4} = \frac{3}{3}$

2007 - 2017 © Frank Ho, Amanda Ho, All rights reserved.　　www.homathchess.com

Different ways of writing multiplication (Learning division while doing multiplications)

Ho Math Chess　　何数棋谜　妈！我会棋谜式乘法啦！
Mom! I Learn Multiplication Using Math-Chess-Puzzles Connection!

Student's Name _____ Date _____

2007 - 2017 © Frank Ho, Amanda Ho, All rights reserved.　　www.homathchess.com

Different ways of writing multiplication (Learning division while doing multiplications)

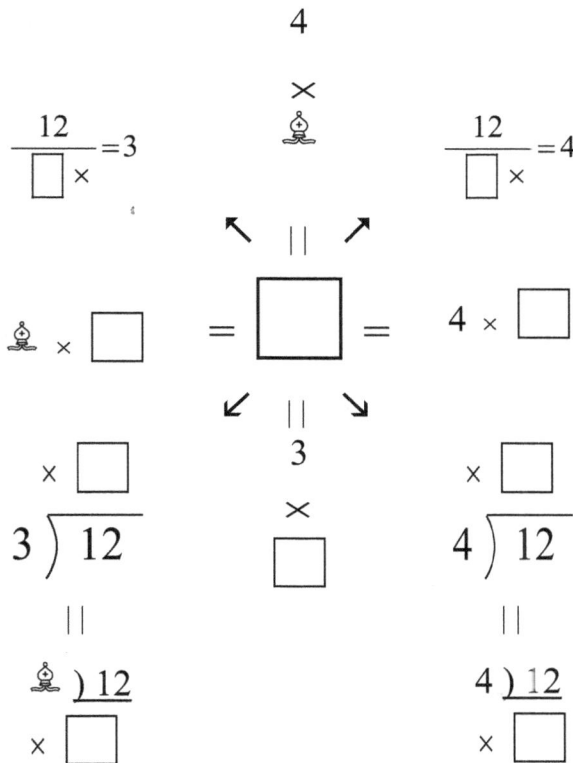

2007 - 2017 © Frank Ho, Amanda Ho, All rights reserved.   www.homathchess.com

Different ways of writing multiplication (Learning division while doing multiplications)

$$\frac{16}{\square \times} = 4$$

$$\begin{array}{c} 4 \\ \times \\ 4 \end{array}$$

$$\frac{16}{\square \times} = 4$$

$$4 \times \square \quad = \quad \boxed{} \quad = \quad \square \times 4$$

$$\begin{array}{c} \times \square \\ \hline 4 \overline{)16} \end{array}$$

$$\begin{array}{c} 4 \\ \times \\ \square \end{array}$$

$$\begin{array}{c} \times \square \\ \hline 4 \overline{)16} \end{array}$$

$$\begin{array}{c} 4 \overline{)16} \\ \times \square \end{array}$$

$$\begin{array}{c} 4 \overline{)16} \\ \times \square \end{array}$$

Different ways of writing multiplication (Learning division while doing multiplications)

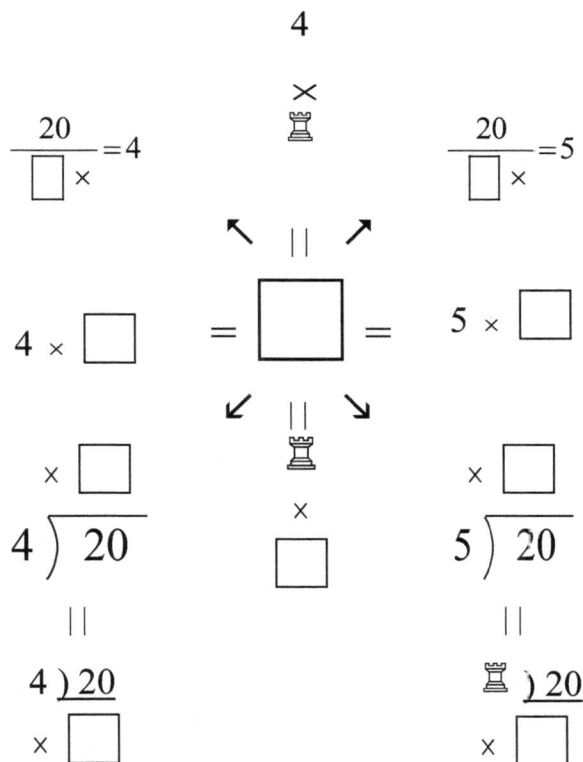

$$\frac{20}{\boxed{} \times} = 4 \qquad \frac{4}{\times \;\text{♖}} \qquad \frac{20}{\boxed{} \times} = 5$$

$$4 \times \boxed{} \quad = \quad \boxed{} \quad = \quad 5 \times \boxed{}$$

$$\begin{array}{r} \times \;\boxed{} \\ 4 \,\overline{\smash{)}\,20} \end{array} \qquad \begin{array}{c} \text{♖} \\ \times \\ \boxed{} \end{array} \qquad \begin{array}{r} \times \;\boxed{} \\ 5 \,\overline{\smash{)}\,20} \end{array}$$

$$\begin{array}{c} 4 \,\overline{)\,20} \\ \times \;\boxed{} \end{array} \qquad\qquad \begin{array}{c} \text{♖} \,\overline{)\,20} \\ \times \;\boxed{} \end{array}$$

2007 - 2017 © Frank Ho, Amanda Ho, All rights reserved.   www.homathchess.com

Different ways of writing multiplication (Learning division while doing multiplications)

$$\begin{array}{c} 4 \\ \times \\ 6 \end{array}$$

$$\frac{24}{\square \times} = 4 \qquad\qquad \frac{24}{\square \times} = 6$$

$$4 \times \square \quad = \boxed{} = \quad 6 \times \square$$

$$\times \, \square \qquad\qquad 6 \qquad\qquad \times \, \square$$

$$4\,\overline{)\,24} \qquad \begin{array}{c} \times \\ \square \end{array} \qquad 6\,\overline{)\,24}$$

$$4\,\underline{)\,24} \qquad\qquad 6\,\underline{)\,24}$$
$$\times \, \square \qquad\qquad\qquad \times \, \square$$

Different ways of writing multiplication (Learning division while doing multiplications)

$$\frac{28}{\square \times} = 4 \qquad \frac{4}{\times 7} \qquad \frac{28}{\square \times} = 7$$

$$4 \times \square \quad = \quad \boxed{} \quad = \quad 7 \times \square$$

$$\times \square \qquad \qquad \times \square$$

$$4\overline{)28} \qquad \frac{7}{\times \square} \qquad 7\overline{)28}$$

$$4\,\underline{)\,28} \qquad\qquad 7\,\underline{)\,28}$$
$$\times \square \qquad\qquad\qquad \times \square$$

2007 - 2017 © Frank Ho, Amanda Ho, All rights reserved.   www.homathchess.com

Different ways of writing multiplication (Learning division while doing multiplications)

$$\frac{32}{\Box \times} = 4 \qquad \begin{array}{c} 4 \\ \times \\ 8 \end{array} \qquad \frac{32}{\Box \times} = 8$$

$$4 \times \Box \quad = \boxed{} = \quad 8 \times \Box$$

$$\begin{array}{c} \times \Box \\ 4\,\overline{)\,32} \\ || \\ 4\,\underline{)\,32} \\ \times \Box \end{array} \qquad \begin{array}{c} 8 \\ \times \\ \Box \end{array} \qquad \begin{array}{c} \times \Box \\ 8\,\overline{)\,32} \\ || \\ 8\,\underline{)\,32} \\ \times \Box \end{array}$$

Ho Math Chess 何数棋谜 妈！我会棋谜式乘法啦！
Mom! I Learn Multiplication Using Math-Chess-Puzzles Connection!

Student's Name _____ Date _____

2007 - 2017 © Frank Ho, Amanda Ho, All rights reserved.　　www.homathchess.com

Different ways of writing multiplication (Learning division while doing multiplications)

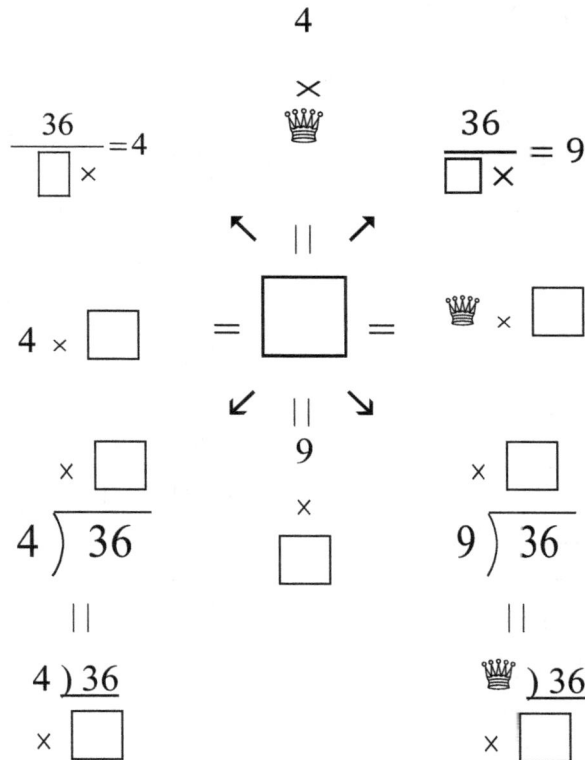

2007 - 2017 © Frank Ho, Amanda Ho, All rights reserved.　　www.homathchess.com

Counting 5's multiples

Write 5's multiples.

♜, 10, 15, ☐,☐,☐,☐,☐,☐

Fill in the following ☐ with a number.

Sequence	♙	2	♗	4	♜	6	7	8	♛
Add 5	☐	10	☐	20	☐	30	☐	40	☐

Sequence	1	2	3	4	5	6	7	8	♛
Add 5	♜	☐	15	☐	25	☐	35	☐	45

Sequence	♙	2	♗	4	5	6	7	8	9
Add 5	☐	10	☐	20	☐	30	☐	40	☐

Sequence	1	2	3	4	5	6	7	8	♛
Add 5	5	☐	15	☐	25	☐	35	☐	45

Sequence	♙	2	3	4	♜	6	7	8	9
Add 5	☐	10	☐	20	☐	30	☐	40	☐

Sequence	1	2	♗	4	5	6	7	8	♛
Add 5	5	☐	15	☐	25	☐	35	☐	45

2007 - 2017 © Frank Ho, Amanda Ho, All rights reserved.　　www.homathchess.com

5 times

5 × 2 = ☐	Five times two is ☐	2 × 5 = ☐	two times five is ☐
5 × 3 = ☐	Five times three is ☐	3 × ♖ = ☐	Three times five is ☐
5 × 4 = ☐	Five times four is ☐	4 × 5 = ☐	Four times five is ☐
♖ × 5 = ☐	Five times five is ☐	5 × ♖ = ☐	Five times five is ☐
5 × 6 = ☐	Five times six is ☐	6 × 5 = ☐	Six times five is ☐
♖ × 7 = ☐	Five times seven is ☐	7 × 5 = ☐	Seven times five is ☐
5 × 8 = ☐	Five times eight is ☐	8 × ♖ = ☐	Eight times five is ☐
♖ × 9 = ☐	Five times nine is ☐	9 × 5 = ☐	Nine times five is ☐

♖	1	2	♖	♘
× 1	× 5	× 5	× 2	× 5
☐	☐	☐☐	☐☐	☐☐

5	5	♖	5	5
× 5	× 6	× 7	× 8	× 9
☐☐	☐☐	☐☐	☐☐	☐☐

4	♘	5	7	5
× 5	× ♖	× 5	× 5	× ♛
☐☐	☐☐	☐☐	☐☐	☐☐

Ho Math Chess 何数棋谜 妈！我会棋谜式乘法啦！

Mom! I Learn Multiplication Using Math-Chess-Puzzles Connection!

Student's Name _____ Date _____

2007 - 2017 © Frank Ho, Amanda Ho, All rights reserved. www.homathchess.com

♖	1	5	2	5
X 1	X 5	X 2	X 5	X 3

5	5	7	♖	9
X 4	X 5	X 5	X 8	X 5

5	♖	5	5	♕
X 6	X 8	X 7	X 6	X ♖

♖	5	7	♖	♕
X ♖	X 6	X ♖	X 8	X 5

7	5	♖	5	5
X ♖	X 8	X 2	X 6	X 9

Mom! I Learn Multiplication Using Math-Chess-Puzzles Connection!

Student's Name _____ Date _____

2007 - 2017 © Frank Ho, Amanda Ho, All rights reserved.   www.homathchess.com

Oral practice

five one five	5 1 ☐	5 1 × ♖ 255 ☐☐☐
five two ten	♖ 2 ☐	5 2 × 5 260 ☐☐☐
five three fifteen	5 3 ☐	5 3 × ♖ 265 ☐☐☐
five four twenty	5 4 ☐	5 4 × 5 270 ☐☐☐
five five twenty-five	♖ 5 ☐	$\overset{2}{5}5$ × 5 275 ☐☐☐

2007 - 2017 © Frank Ho, Amanda Ho, All rights reserved.　www.homathchess.com

Oral practice

five six thirty	♖ 6 ☐	^{3} 56 $\times\ 5$ $280\ ☐☐☐$
five seven thirty-five	5 7 ☐	57 $\times\ ♖$ $285\ ☐☐☐$
five eight forty	♖ 8 ☐	58 $\times\ 5$ $290\ ☐☐☐$
five nine forty-five	5 9 ☐	59 $\times\ 5$ $295\ ☐☐☐$
five five twenty-five	♖ 5 ☐	55 $\times\ 5$ $275\ ☐☐☐$
five six thirty	5 6 ☐	56 $\times\ ♖$ $280\ ☐☐☐$

Fill in ☐ with answer.

Times	Grouping	Addition
$5 \times \$1 = \square$	5 of $\square = 5$	$\$1 + \$1 + \$1 + \$1 + \$1 = \square$
$1 \times \$5 = \square$	1 of $\square = 5$	$\$5$

Fill in ☐ with answer.

Expression	Grouping	Addition
$5 \times \$2$	5 of $\square = 10$	$\$2 + \$2 + \$2 + \$2 + \$2 = \square$
$2 \times \$5$	2 of $\square = 10$	$\$5 + \$5 = \square$

$5 \times ♟ = \square = ♟ \times 5 = \square$	$♟ \times ♜ = \square = 5 \times 1 = \square$
$5 \times \square = 10 = 2 \times \square = \square$	$2 \times \square = 10 = 5 \times \square = \square$

5 1 ☐	5 5 ☐	5 9 ☐	5 4 ☐	5 8 ☐
♜ 2 ☐	5 6 ☐	♜ ♟ ☐	5 5 ☐	5 9 ☐
5 3 ☐	5 7 ☐	5 2 ☐	5 6 ☐	5 1 ☐
5 4 ☐	5 8 ☐	5 3 ☐	5 7 ☐	♜ 2 ☐

2007 - 2017 © Frank Ho, Amanda Ho, All rights reserved.   www.homathchess.com

Fill in ☐ with answer.

Times	Grouping	Addition
$5 \times \$3 = \square$	$5 \text{ of } \square = 15$	$\$3 + \$3 + \$3 + \$3 + \$3 = \square$
$3 \times \$5 = \square$	$3 \text{ of } \square = 15$	$\$5 + \$5 + \$5 = \square$

Fill in ☐ with answer.

Expression	Grouping	Addition
$5 \times \$4$	$5 \text{ of } \square = 20$	$\$4 + \$4 + \$4 + \$4 + \$4 = \square$
$4 \times \$5$	$4 \text{ of } \square = 20$	$\$5 + \$5 + \$5 + \$5 = \square$

$5 \times 3 = \square = 3 \times 5 = \square$	$3 \times ♖ = \square = 5 \times 3 = \square$
$♖ \times \square = 20 = 4 \times \square = \square$	$4 \times \square = 20 = 5 \times \square = \square$

$5 ♙ \square$	$5\ 5\ \square$	$5\ 9\ \square$	$5\ 4\ \square$	$♖\ 8\ \square$
$5\ 2\ \square$	$5\ 6\ \square$	$♖\ 1\ \square$	$5\ 5\ \square$	$♖\ 9\ \square$
$5\ 3\ \square$	$♖\ 7\ \square$	$5\ 2\ \square$	$♖\ 6\ \square$	$5 ♙ \square$
$♖\ 4\ \square$	$5\ 8\ \square$	$5\ 3\ \square$	$5\ 7\ \square$	$5\ 2\ \square$

Ho Math Chess 何数棋谜 妈！我会棋谜式乘法啦！
Mom! I Learn Multiplication Using Math-Chess-Puzzles Connection!

Student's Name _____ Date _____

2007 - 2017 © Frank Ho, Amanda Ho, All rights reserved.   www.homathchess.com

Fill in ☐ with answer.

Times	Grouping	Addition
$5 \times \$5 =$ ☐	♖ of ☐ $= 25$	$\$5 + \$5 + \$5 + \$5 + \$5 =$ ☐
$5 \times \$5 =$ ☐	5 of ☐ $= 25$	$\$5 + \$5 + \$5 + \$5 + \$5 =$ ☐

Times	Grouping	Addition
$5 \times \$6 =$ ☐	5 of ☐ $= 30$	$\$6 + \$6 + \$6 + \$6 + \$6 =$ ☐
$6 \times \$5 =$ ☐	6 of ☐ $= 30$	$\$5 + \$5 + \$5 + \$5 + \$5 + \$5 =$ ☐

$5 \times 5 =$ ☐ $= 5 \times 5 =$ ☐	$5 \times 5 =$ ☐ $= 5 \times 5 =$ ☐
$5 \times$ ☐ $= 30 = 6 \times$ ☐ $=$ ☐	$6 \times$ ☐ $= 30 = 5 \times$ ☐ $=$ ☐

5 1 ☐	5 5 ☐	5 9 ☐	5 4 ☐	♖ 8 ☐
♖ 2 ☐	5 6 ☐	5 1 ☐	♖ 5 ☐	5 9 ☐
5 3 ☐	♖ 7 ☐	5 2 ☐	5 6 ☐	5 1 ☐
5 4 ☐	5 8 ☐	5 3 ☐	5 7 ☐	♖ 2 ☐

2007 - 2017 © Frank Ho, Amanda Ho, All rights reserved.　　www.homathchess.com

Fill in _____ and ☐ with answers.

Times	Grouping	Addition
$5 \times \$7 = \square$	5 of $\square$ = 35	$\$7 + \$7 + \$7 + \$7 + \$7 = \square$
$7 \times \$5 = \square$	7 of $\square$ = 35	$\$5 + \$5 + \$5 + \$5 + \$5 + \$5 + \$5 = \square$

Times	Grouping	Addition
♖ $\times \$8 = \square$	5 of $\square$ = 40	$\$8 + \$8 + \$8 + \$8 + \$8 = \square$
$8 \times \$♖ = \square$	8 of $\square$ = 40	$\$5 + \$5 + \$5 + \$5 + \$5 + \$5 + \$5 + \$5 = \square$

$5 \times 7 = \square = 7 \times 5 = \square$	$7 \times 5 = \square = 5 \times 7 = \square$
$5 \times \square = 40 = 8 \times \square = \square$	$8 \times \square = 40 = 5 \times \square = \square$

5 1 ☐	5 5 ☐	♖ 9 ☐	5 4 ☐	♖ 8 ☐
♖ 2 ☐	5 6 ☐	5 1 ☐	5 5 ☐	5 9 ☐
5 3 ☐	♖ 7 ☐	5 2 ☐	♖ 6 ☐	5 1 ☐
5 4 ☐	5 8 ☐	5 3 ☐	5 7 ☐	♖ 2 ☐

Fill in _____ and ☐ with answers.

Times	Grouping	Addition
$5 \times \$9 = \square$	♖ of $\square$ = 45	$\$9 + \$9 + \$9 + \$9 + \$9 = \square$
$9 \times \$5 = \square$	9 of $\square$ = 45	$\$5 + \$5 + \$5 + \$5 + \$5 + \$5 + \$5 + \$5 + \$5 = \square$

Times	Grouping	Addition
♖ $\times \$8 = \square$	5 of $\square$ = 40	$\$8 + \$8 + \$8 + \$8 + \$8 = \square$
$8 \times \$5 = \square$	8 of $\square$ = 40	$\$5 + \$5 + \$5 + \$5 + \$5 + \$5 + \$5 + \$5 = \square$

$5 \times 9 = \square = 9 \times ♖ = \square$	$9 \times ♖ = \square = 5 \times 9 = \square$
$5 \times \square = 40 = 8 \times \square = \square$	$8 \times \square = 40 = 5 \times \square = \square$

5 1 ☐	5 5 ☐	5 9 ☐	5 4 ☐	♖ 8 ☐
5 2 ☐	♖ 6 ☐	5 1 ☐	♖ 5 ☐	5 9 ☐
5 3 ☐	5 7 ☐	♖ 2 ☐	5 6 ☐	5 1 ☐
♖ 4 ☐	5 8 ☐	5 3 ☐	5 7 ☐	♖ 2 ☐

Ho Math Chess   何数棋谜　妈！我会棋谜式乘法啦！
Mom! I Learn Multiplication Using Math-Chess-Puzzles Connection!
Student's Name _____ Date _____
2007 - 2017 © Frank Ho, Amanda Ho, All rights reserved.   www.homathchess.com

Preparing for division

□	□	□	□	□
X 4	X 2	X 6	X 9	X 3
20	10	30	45	15

□	□	□	□	□
X 8	X 6	X ♜	X 3	X 2
40	30	25	15	10

□	□	□	□	□
X 9	X ♜	X 7	X 5	X 7
45	25	35	25	35

□	□	□	□	□
X 8	X 6	X ♜	X 3	X 2
40	30	25	15	10

□	□	□	□	□
X ♛	X 6	X 7	X ♜	X 7
45	30	35	30	35

2007 - 2017 © Frank Ho, Amanda Ho, All rights reserved.　　www.homathchess.com

Preparing for division

□ X ♖ = 5	X □ 5)5	□)5 X ♖
□ X 5 = 10	X □ 5)10	□)10 X 5
□ X ♖ = 15	X □ 5)15	□)15 X ♖
□ X 5 = 20	X □ 5)20	□)20 X 5
□ X ♖ = 25	X □ 5)25	□)25 X 5
□ X 5 = 30	X □ 5)30	□)30 X ♖
□ X 5 = 35	X □ 5)35	□)35 X 5

2007 - 2017 © Frank Ho, Amanda Ho, All rights reserved.   www.homathchess.com

Preparing for division

☐ X 5 = 40	X ☐ 5)40	☐)40 X 5
☐ X 5 = 45	X ☐ 5)45	☐)45 X 5
☐ X ♖ = 5	X ☐ 5)5	☐)5 X ♖
☐ X 5 = 10	X ☐ 5)10	☐)10 X 5
☐ X ♖ = 15	X ☐ 5)15	☐)15 X 5
☐ X 5 = 20	X ☐ 5)20	☐)20 X ♖
☐ X ♖ = 25	X ☐ 5)25	☐)25 X 5

Cross multiplication

12 $\nwarrow$ 12 $\nearrow$ $\frac{6}{2} = \frac{6}{2}$	☐ **10** $\nwarrow$ $\nearrow$ $\frac{5}{5} = \frac{2}{2}$	☐ **15** $\nwarrow$ $\nearrow$ $\frac{5}{5} = \frac{3}{3}$	☐ **20** $\nwarrow$ $\nearrow$ $\frac{5}{5} = \frac{4}{4}$
☐ **25** $\nwarrow$ $\nearrow$ $\frac{5}{5} = \frac{5}{5}$	☐ **30** $\nwarrow$ $\nearrow$ $\frac{5}{5} = \frac{6}{6}$	☐ **35** $\nwarrow$ $\nearrow$ $\frac{5}{5} = \frac{7}{7}$	☐ **45** $\nwarrow$ $\nearrow$ $\frac{5}{5} = \frac{9}{9}$
☐ **25** $\nwarrow$ $\nearrow$ $\frac{5}{5} = \frac{5}{5}$	☐ **20** $\nwarrow$ $\nearrow$ $\frac{5}{5} = \frac{4}{4}$	☐ **35** $\nwarrow$ $\nearrow$ $\frac{5}{5} = \frac{7}{7}$	☐ **40** $\nwarrow$ $\nearrow$ $\frac{5}{5} = \frac{8}{8}$
☐ **30** $\nwarrow$ $\nearrow$ $\frac{5}{5} = \frac{6}{6}$	☐ **40** $\nwarrow$ $\nearrow$ $\frac{5}{5} = \frac{8}{8}$	☐ **45** $\nwarrow$ $\nearrow$ $\frac{5}{5} = \frac{9}{9}$	☐ **15** $\nwarrow$ $\nearrow$ $\frac{5}{5} = \frac{3}{3}$

Different ways of writing multiplication (Learning division while doing multiplications)

♖

×

2

$$\frac{10}{\boxed{}\times} = 2 \qquad\qquad \frac{10}{\boxed{}\times} = 2$$

↖ ‖ ↗

$$2 \times \boxed{} \quad = \quad \boxed{} \quad = \quad ♖ \times \boxed{}$$

↙ ‖ ↘

$$\times \boxed{} \qquad\qquad 2 \qquad\qquad \times \boxed{}$$

$$2\,\overline{)\,10} \qquad\qquad × \qquad\qquad 5\,\overline{)\,10}$$

$$\boxed{}$$

‖ ‖

$$2\,)\,10 \qquad\qquad\qquad ♖\,)\,10$$

$$× \boxed{} \qquad\qquad\qquad × \boxed{}$$

Different ways of writing multiplication (Learning division while doing multiplications)

Ho Math Chess 何数棋谜 妈!我会棋谜式乘法啦!
Mom! I Learn Multiplication Using Math-Chess-Puzzles Connection!

Student's Name _____ Date _____

2007 - 2017 © Frank Ho, Amanda Ho, All rights reserved.   www.homathchess.com

Different ways of writing multiplication (Learning division while doing multiplications)

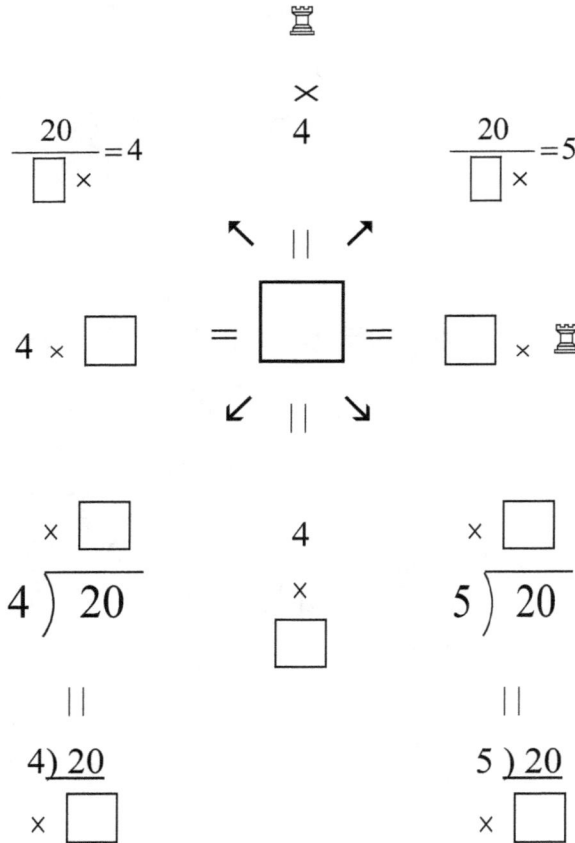

$$\frac{20}{\boxed{} \times} = 4 \qquad \qquad \frac{20}{\boxed{} \times} = 5$$

$$4 \times \boxed{} = \boxed{} = \boxed{} \times$$

Ho Math Chess 何数棋谜 妈！我会棋谜式乘法啦！
Mom! I Learn Multiplication Using Math-Chess-Puzzles Connection!

Student's Name _____ Date _____

2007 - 2017 © Frank Ho, Amanda Ho, All rights reserved.   www.homathchess.com

Different ways of writing multiplication (Learning division while doing multiplications)

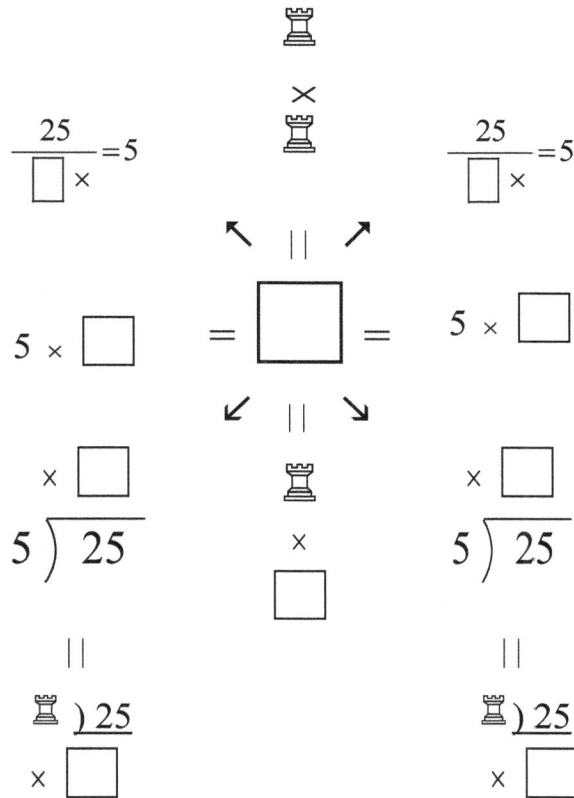

$$\frac{25}{\square_\times} = 5 \qquad\qquad \frac{25}{\square_\times} = 5$$

$$5 \times \square \quad = \quad \boxed{} \quad = \quad 5 \times \square$$

$$\begin{array}{r} \times \square \\ 5\overline{)\,25} \end{array} \qquad\qquad \begin{array}{r} \times \square \\ 5\overline{)\,25} \end{array}$$

$$\begin{array}{r} \text{♖}\,\overline{)\,25} \\ \times \square \end{array} \qquad\qquad \begin{array}{r} \text{♖}\,\overline{)\,25} \\ \times \square \end{array}$$

2007 - 2017 © Frank Ho, Amanda Ho, All rights reserved.   www.homathchess.com

Different ways of writing multiplication (Learning division while doing multiplications)

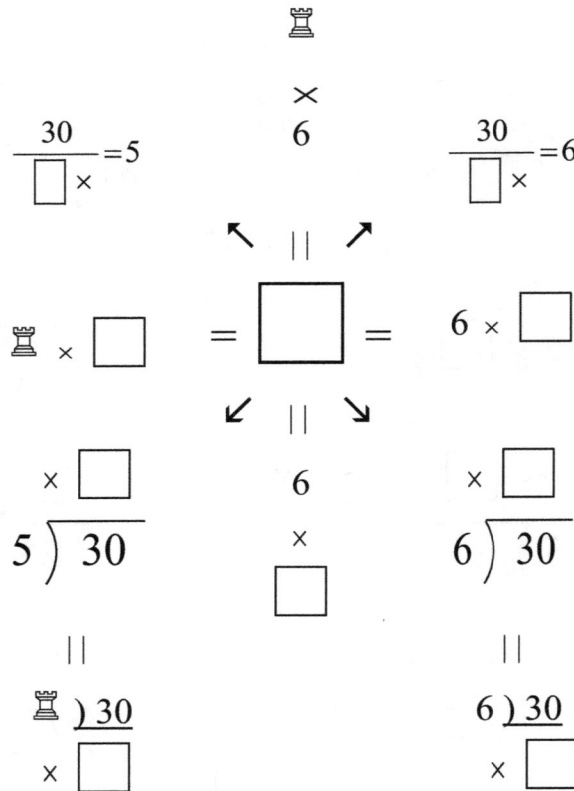

$$\frac{30}{\square\,\times} = 5 \qquad \frac{30}{\square\,\times} = 6$$

$$\square \times 6$$

$$\unicode{x265C} \times \square \quad = \quad \boxed{} \quad = \quad 6 \times \square$$

$$\times\ \square$$
$$5\,\overline{)\,30} \qquad 6 \times \boxed{} \qquad 6\,\overline{)\,30}\ \times\ \square$$

$$\unicode{x265C}\,)\,30 \qquad\qquad 6\,)\,30$$
$$\times\ \square \qquad\qquad\quad \times\ \square$$

2007 - 2017 © Frank Ho, Amanda Ho, All rights reserved.   www.homathchess.com

Different ways of writing multiplication (Learning division while doing multiplications)

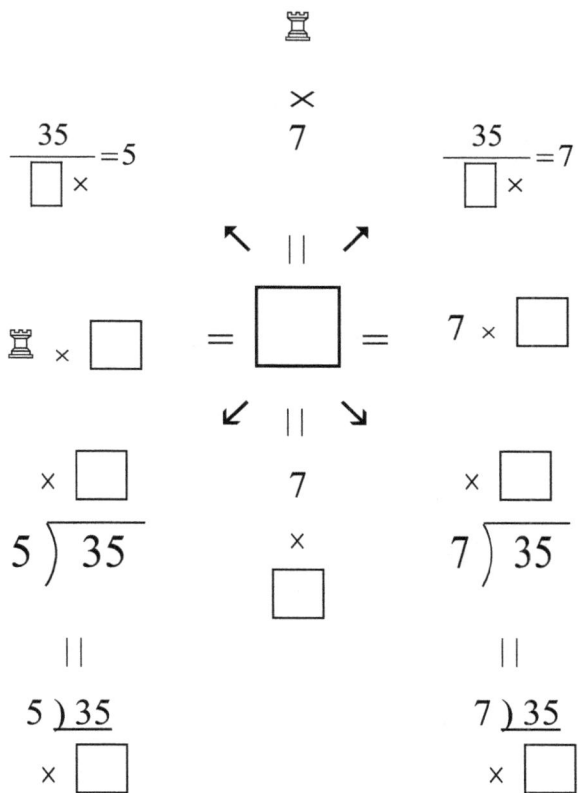

$$\frac{35}{\Box \times} = 5 \qquad \times \atop 7 \qquad \frac{35}{\Box \times} = 7$$

$$\Box \times \Box = \boxed{} = 7 \times \Box$$

$$5\overline{)35} \qquad \times \atop 7 \times \Box \qquad 7\overline{)35}$$

$$5\underline{)35} \atop \times \Box \qquad 7\underline{)35} \atop \times \Box$$

2007 - 2017 © Frank Ho, Amanda Ho, All rights reserved.　　www.homathchess.com

Different ways of writing multiplication (Learning division while doing multiplications)

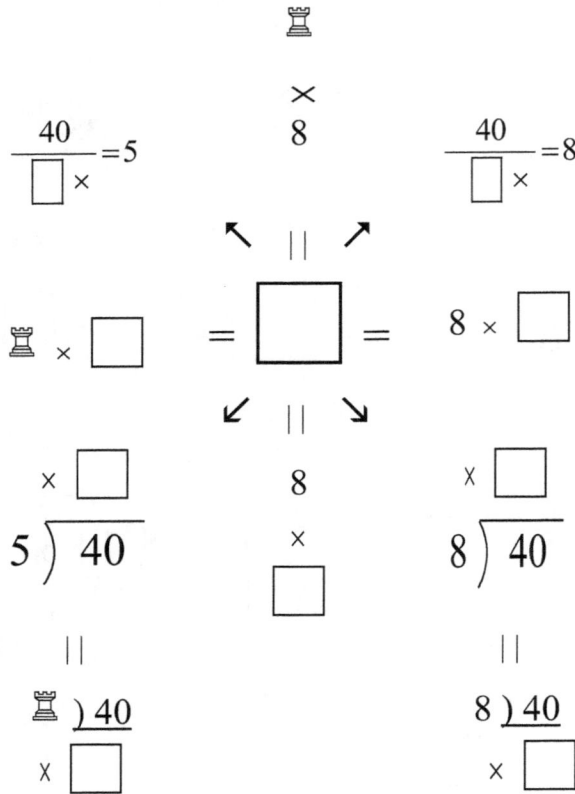

$$\frac{40}{\square\times}=5 \qquad \overset{\text{♜}}{\underset{8}{\times}} \qquad \frac{40}{\square\times}=8$$

$$\text{♜}\times\square \;=\; \boxed{} \;=\; 8\times\square$$

$$5\overline{)\,40} \qquad \overset{8}{\underset{\square}{\times}} \qquad 8\overline{)\,40}$$

$$\text{♜}\,)\,40 \qquad\qquad 8\,)\,40$$
$$\times\,\square \qquad\qquad\qquad \times\,\square$$

Ho Math Chess 何数棋谜 妈！我会棋谜式乘法啦！
Mom! I Learn Multiplication Using Math-Chess-Puzzles Connection!

Student's Name _____ Date _____

2007 - 2017 © Frank Ho, Amanda Ho, All rights reserved.   www.homathchess.com

Different ways of writing multiplication (Learning division while doing multiplications)

Ho Math Chess   何数棋谜   妈!我会棋谜式乘法啦!
Mom! I Learn Multiplication Using Math-Chess-Puzzles Connection!

Student's Name _____ Date _____

2007 - 2017 © Frank Ho, Amanda Ho, All rights reserved.   www.homathchess.com

Counting 6s multiples

6, 12, 18, ☐,☐,☐,☐,☐,☐

Fill in the following ☐ with a number.

Sequence	♙	2	3	4	5	6	7	8	♛
Add 6	☐	12	☐	24	☐	36	☐	48	☐

Sequence	♙	2	3	4	♜	6	7	8	9
Add 6	6	☐	18	☐	30	☐	42	☐	54

Sequence	1	2	3	4	♜	6	7	8	♛
Add 6	☐	12	☐	24	☐	36	☐	48	☐

Sequence	♙	2	♝	4	5	6	7	8	9
Add 6	6	☐	18	☐	30	☐	42	☐	54

Sequence	1	2	♝	4	5	6	7	8	♛
Add 6	☐	12	☐	24	☐	36	☐	48	☐

Sequence	♙	2	3	4	♖	6	7	8	9
Add 6	6	☐	18	☐	30	☐	42	☐	54

2007 - 2017 © Frank Ho, Amanda Ho, All rights reserved.　www.homathchess.com

6 times

6 × 1 = ☐	Six times one is ☐	♟ × 6= ☐	One times six is ☐
6 × 2 = ☐	Six times two is ☐	2 × 6 = ☐	Two times six is ☐
6 × 3 = ☐	Six times three is ☐	3 × 6 = ☐	Three times six is ☐
6 × 4 = ☐	Six times four is ☐	4 × 6 = ☐	Four times six is ☐
6 × 5 = ☐	Six times five is ☐	5 × 6 = ☐	Five times six is ☐
6 × 6 = ☐	Six times six is ☐	6 × 6 = ☐	Six times six is ☐
6 × 7 = ☐	Six times seven is ☐	7 × 6 = ☐	Seven times six is ☐
6 × 8 = ☐	Six times eight is ☐	8 × 6 = ☐	Eight times six is ☐
6 × 9 = ☐	Six times nine is ☐	9 × 6 = ☐	Nine times six is ☐

6	♟	2	6	♗
X 1	X 6	X 6	X 2	X 6
☐	☐	☐☐	☐☐	☐☐

5	6	6	8	5
X 6	X 6	X 7	X 6	X 3
☐☐	☐☐	☐☐	☐☐	☐☐

4	♗	♖	7	6
X 6	X 6	X 6	X 6	X 9
☐☐	☐☐	☐☐	☐☐	☐☐

2007 - 2017 © Frank Ho, Amanda Ho, All rights reserved.　　www.homathchess.com

6	1	6	2	6
X 1	X 6	X 2	X 6	X 3

6	6	7	6	♛
X 4	X 5	X 6	X 8	X 6

6	6	6	6	9
X 6	X 8	X 7	X 6	X 6

6	6	7	6	♛
X ♜	X 6	X 6	X 8	X 6

7	6	5	6	6
X 6	X 8	X 6	X 6	X 9

Ho Math Chess 何数棋谜 妈！我会棋谜式乘法啦！

Mom! I Learn Multiplication Using Math-Chess-Puzzles Connection!

Student's Name _____ Date _____

2007 - 2017 © Frank Ho, Amanda Ho, All rights reserved.   www.homathchess.com

Oral practice

six one six	6 1 ☐	$\begin{array}{r}1\ 1\\ \times\quad 6\\ \hline 66\ \square\square\end{array}$
six two twelve	6 2 ☐	$\begin{array}{r}2\ 2\\ \times\quad 6\\ \hline 132\square\square\square\end{array}$
six three eighteen	6 ♞ ☐	$\begin{array}{r}3\ 3\\ \times\quad 6\\ \hline 198\square\square\square\end{array}$
six four twenty-four	6 4 ☐	$\begin{array}{r}4\ 4\\ \times\quad 6\\ \hline 264\square\square\square\end{array}$
six five thirty	6 5 ☐	$\begin{array}{r}^{3}\\ 5\ 5\\ \times\quad 6\\ \hline 330\square\square\square\end{array}$

2007 - 2017 © Frank Ho, Amanda Ho, All rights reserved.   www.homathchess.com

Oral practice

six six thirty-six	6 6 ☐	$\begin{array}{r} {}^{3} \\ 6\,6 \\ \times\quad 6 \\ \hline 396\,\square\square\square \end{array}$
six seven forty-two	6 7 ☐	$\begin{array}{r} 7\,7 \\ \times\quad 6 \\ \hline 462\,\square\square\square \end{array}$
six eight forty-eight	6 8 ☐	$\begin{array}{r} 8\,8 \\ \times\quad 6 \\ \hline 528\,\square\square\square \end{array}$
six nine fifty-four	6 ♛ ☐	$\begin{array}{r} 9\,9 \\ \times\quad 6 \\ \hline 594\,\square\square\square \end{array}$
six five thirty	6 ♜ ☐	$\begin{array}{r} 5\,5 \\ \times\quad 6 \\ \hline 330\,\square\square\square \end{array}$
six six thirty-six	6 6 ☐	$\begin{array}{r} 6\,6 \\ \times\quad 6 \\ \hline 396\,\square\square\square \end{array}$

Ho Math Chess 何数棋谜 妈！我会棋谜式乘法啦！

Mom! I Learn Multiplication Using Math-Chess-Puzzles Connection!

Student's Name _____ Date _____

2007 - 2017 © Frank Ho, Amanda Ho, All rights reserved.   www.homathchess.com

Fill in ☐ with answer.

Times	Grouping	Addition
$6 \times \$1 = $ ☐	6 of ☐ $= 6$	$\$1 + \$1 + \$1 + \$1 + \$1 + \$1 = $ ☐
$1 \times \$6 = $ ☐	1 of ☐ $= 6$	$\$6$

Fill in ☐ with answer.

Expression	Grouping	Addition
$6 \times \$2$	6 of ☐ $= 12$	$\$2 + \$2 + \$2 + \$2 + \$2 + \$2 = $ ☐
$2 \times \$6$	2 of ☐ $= 12$	$\$6 + \$6 = $ ☐

$6 \times 1 = $ ☐ $= 1 \times 6 = $ ☐	$1 \times 6 = $ ☐ $= 6 \times 1 = $ ☐
$6 \times $ ☐ $= 12 = 2 \times $ ☐ $= $ ☐	$2 \times $ ☐ $= 12 = 6 \times $ ☐ $= $ ☐

6 ♙ ☐	6 5 ☐	6 9 ☐	6 4 ☐	6 8 ☐
6 2 ☐	6 6 ☐	6 1 ☐	6 ♖ ☐	6 ♕ ☐
6 ♗ ☐	6 7 ☐	6 2 ☐	6 6 ☐	6 1 ☐
6 4 ☐	6 8 ☐	6 ♗ ☐	6 7 ☐	6 2 ☐

2007 - 2017 © Frank Ho, Amanda Ho, All rights reserved.   www.homathchess.com

Fill in ☐ with answer.

Times	Grouping	Addition
6 × \$♟ = ☐	6 of ☐ = 18	\$3 + \$3 + \$3 + \$3 + \$3 + \$3 = ☐
3 × \$6 = ☐	3 of ☐ = 18	\$6 + \$6 + \$6 = ☐

Fill in ☐ with answer.

Expression	Grouping	Addition
6 × \$4	6 of ☐ = 24	\$4 + \$4 + \$4 + \$4 + \$4 = ☐
4 × \$6	4 of ☐ = 24	\$6 + \$6 + \$6 + \$6 = ☐

6 × 3 = ☐ = ♟ × 6 = ☐	3 × 6 = ☐ = 6 × 3 = ☐
6 × ☐ = 24 = 4 × ☐ = ☐	4 × ☐ = 24 = 6 × ☐ = ☐

6 1 ☐	6 5 ☐	6 9 ☐	6 4 ☐	6 8 ☐
6 2 ☐	6 6 ☐	6 1 ☐	6 5 ☐	6 9 ☐
6 ♟ ☐	6 7 ☐	6 2 ☐	6 6 ☐	6 ♙ ☐
6 4 ☐	6 8 ☐	6 3 ☐	6 7 ☐	6 2 ☐

Fill in ☐ with answer.

Times	Grouping	Addition
$6 \times \$5 = $ ☐	6 of ☐ = 30	$\$5 + \$5 + \$5 + \$5 + \$5 + \$5 = $ ☐
$5 \times \$6 = $ ☐	5 of ☐ = 30	$\$6 + \$6 + \$6 + \$6 + \$6 = $ ☐

Times	Grouping	Addition
$6 \times \$6 = $ ☐	6 of ☐ = 36	$\$6 + \$6 + \$6 + \$6 + \$6 + \$6 = $ ☐
$6 \times \$6 = $ ☐	6 of ☐ = 36	$\$6 + \$6 + \$6 + \$6 + \$6 + \$6 = $ ☐

$5 \times 6 = $ ☐ $ = 6 \times 5 = $ ☐	$6 \times 5 = $ ☐ $ = 5 \times 6 = $ ☐
$6 \times $ ☐ $ = 36 = 6 \times $ ☐ $ = $ ☐	$6 \times $ ☐ $ = 36 = 6 \times $ ☐ $ = $ ☐

6 1 ☐	6 5 ☐	6 9 ☐	6 4 ☐	6 8 ☐
6 2 ☐	6 6 ☐	6 ♙ ☐	6 5 ☐	6 9 ☐
6 ♗ ☐	6 7 ☐	6 2 ☐	6 6 ☐	6 1 ☐
6 4 ☐	6 8 ☐	6 ♗ ☐	6 7 ☐	6 2 ☐

Ho Math Chess 何数棋谜 妈!我会棋谜式乘法啦!
Mom! I Learn Multiplication Using Math-Chess-Puzzles Connection!

Student's Name _____ Date _____

2007 - 2017 © Frank Ho, Amanda Ho, All rights reserved.   www.homathchess.com

Fill in ☐ with answers.

Times	Grouping	Addition
6 × $7 = ☐	6 of ☐ = 42	$7 + $7 + $7 + $7 + $7 + $7 = ☐
7 × $6 = ☐	7 of ☐ = 42	$6 + $6 + $6 + $6 + $6 + $6 + $6 = ☐

Times	Grouping	Addition
6 × $8 = ☐	6 of ☐ = 48	$8 + $8 + $8 + $8 + $8 + $8 = ☐
8 × $6 = ☐	8 of ☐ = 48	$6 + $6 + $6 + $6 + $6 + $6 + $6 + $6 = ☐

6 × 7 = ☐ = 7 × 6 = ☐	7 × 6 = ☐ = 6 × 7 = ☐
6 × ☐ = 48 = 8 × ☐ = ☐	8 × ☐ = 48 = 6 × ☐ = ☐

6 1 ☐	6 5 ☐	6 9 ☐	6 4 ☐	6 8 ☐
6 2 ☐	6 6 ☐	6 ♟ ☐	6 5 ☐	6 ♛ ☐
6 ♝ ☐	6 7 ☐	6 2 ☐	6 6 ☐	6 1 ☐
6 4 ☐	6 8 ☐	6 3 ☐	6 7 ☐	6 2 ☐

Student's Name _____ Date _____
2007 - 2017 © Frank Ho, Amanda Ho, All rights reserved.   www.homathchess.com

Fill in ☐ with answer.

Times	Grouping	Addition
$6 \times \$9 = \Box$	$6 \text{ of } \Box = 54$	$\$9 + \$9 + \$9 + \$9 + \$9 + \$9 = \Box$
$9 \times \$6 = \Box$	$9 \text{ of } \Box = 54$	$\$6 + \$6 + \$6 + \$6 + \$6 + \$6 + \$6 + \$6 + \$6 = \Box$

Times	Grouping	Addition
$6 \times \$8 = \Box$	$6 \text{ of } \Box = 48$	$\$8 + \$8 + \$8 + \$8 + \$8 + \$8 = \Box$
$8 \times \$6 = \Box$	$8 \text{ of } \Box = 48$	$\$6 + \$6 + \$6 + \$6 + \$6 + \$6 + \$6 + \$6 = \Box$

$6 \times 9 = \Box = 9 \times 6 = \Box$	$9 \times 6 = \Box = 6 \times 9 = \Box$
$6 \times \Box = 48 = 8 \times \Box = \Box$	$8 \times \Box = 48 = 6 \times \Box = \Box$

$6 \,♙\, \Box$	$6 \; 5 \; \Box$	$6 \; 9 \; \Box$	$6 \; 4 \; \Box$	$6 \; 8 \; \Box$
$6 \; 2 \; \Box$	$6 \; 6 \; \Box$	$6 \; 1 \; \Box$	$6 \; 5 \; \Box$	$6 \,♛\, \Box$
$6 \; 3 \; \Box$	$6 \; 7 \; \Box$	$6 \; 2 \; \Box$	$6 \; 6 \; \Box$	$6 \; 1 \; \Box$
$6 \; 4 \; \Box$	$6 \; 8 \; \Box$	$6 \,♝\, \Box$	$6 \; 7 \; \Box$	$6 \; 2 \; \Box$

2007 - 2017 © Frank Ho, Amanda Ho, All rights reserved.   www.homathchess.com

Preparing for division

□	□	□	□	□
X ♗	X 2	X 8	X 9	X 5
18	12	48	54	30

□	□	□	□	□
X ♖	X 6	X 4	X 8	X 2
30	18	24	48	12

□	□	□	□	□
X 4	X 6	X 5	X 6	X 7
24	42	30	36	42

□	□	□	□	□
X 4	X ♖	X 9	X 8	X 7
24	30	54	48	42

□	□	□	□	□
X 4	X 5	X 3	X ♕	X 6
24	30	18	54	36

Ho Math Chess 何数棋谜 妈！我会棋谜式乘法啦！
Mom! I Learn Multiplication Using Math-Chess-Puzzles Connection!

Student's Name _____ Date _____

2007 - 2017 © Frank Ho, Amanda Ho, All rights reserved.   www.homathchess.com

Preparing for division

☐ X 6 = 6	X ☐ 6)6	☐)6 X 6
☐ X 6 = 12	X ☐ 6)12	☐)12 X 6
☐ X 6 = 18	X ☐ 6)18	☐)18 X 6
☐ X 6 = 24	X ☐ 6)24	☐)24 X 6
☐ X 6 = 30	X ☐ 6)30	☐)30 X 6
☐ X 6 = 36	X ☐ 6)36	☐)36 X 6
☐ X 6 = 42	X ☐ 6)42	☐)42 X 6

Preparing for division

☐ X 6 = 48	X ☐ 6)48	☐)48 X 6
☐ X 6 = 54	X ☐ 6)54	☐)54 X 6
☐ X 6 = 30	X ☐ 6)30	☐)30 X 6
☐ X 6 = 36	X ☐ 6)36	☐)36 X 6
☐ X 6 = 42	X ☐ 6)42	☐)42 X 6
☐ X 6 = 48	X ☐ 6)48	☐)48 X 6
☐ X 6 = 54	X ☐ 6)54	☐)54 X 6

120

Cross multiplication

12　　　12 ↖ ↗ $\frac{6}{2} = \frac{6}{2}$	□　　**12** ↖ ↗ $\frac{6}{6} = \frac{2}{2}$	□　　**18** ↖ ↗ $\frac{6}{6} = \frac{3}{3}$	□　　**24** ↖ ↗ $\frac{6}{6} = \frac{4}{4}$
□　　**30** ↖ ↗ $\frac{6}{6} = \frac{5}{5}$	□　　**36** ↖ ↗ $\frac{6}{6} = \frac{6}{6}$	□　　**42** ↖ ↗ $\frac{6}{6} = \frac{7}{7}$	□　　**54** ↖ ↗ $\frac{6}{6} = \frac{9}{9}$
□　　**30** ↖ ↗ $\frac{6}{6} = \frac{5}{5}$	□　　**24** ↖ ↗ $\frac{6}{6} = \frac{4}{4}$	□　　**42** ↖ ↗ $\frac{6}{6} = \frac{7}{7}$	□　　**48** ↖ ↗ $\frac{6}{6} = \frac{8}{8}$
□　　**36** ↖ ↗ $\frac{6}{6} = \frac{6}{6}$	□　　**48** ↖ ↗ $\frac{6}{6} = \frac{8}{8}$	□　　**54** ↖ ↗ $\frac{6}{6} = \frac{9}{9}$	□　　**18** ↖ ↗ $\frac{6}{6} = \frac{3}{3}$

2007 - 2017 © Frank Ho, Amanda Ho, All rights reserved.   www.homathchess.com

Different ways of writing multiplication (Learning division while doing multiplications)

$$6$$
$$\times$$
$$2$$

$$\frac{12}{\Box \times} = 2 \qquad\qquad \frac{12}{\Box \times} = 6$$

$$\nwarrow \ || \ \nearrow$$

$$2 \times \Box \quad = \quad \boxed{} \quad = \quad 6 \times \Box$$

$$\swarrow \ || \ \searrow$$

$$\times \Box \qquad\qquad \times \Box$$

$$2 \overline{)\,12} \qquad\qquad\qquad 6 \overline{)\,12}$$

$$\begin{array}{c} 2 \\ \times \\ \Box \end{array}$$

$$|| \qquad\qquad\qquad\qquad ||$$

$$2 \underline{)\,12} \qquad\qquad\qquad 6 \underline{)\,12}$$
$$\times \Box \qquad\qquad\qquad \times \Box$$

Ho Math Chess 何数棋谜 妈！我会棋谜式乘法啦！
Mom! I Learn Multiplication Using Math-Chess-Puzzles Connection!

Student's Name _____ Date _____

2007 - 2017 © Frank Ho, Amanda Ho, All rights reserved.   www.homathchess.com

Different ways of writing multiplication (Learning division while doing multiplications)

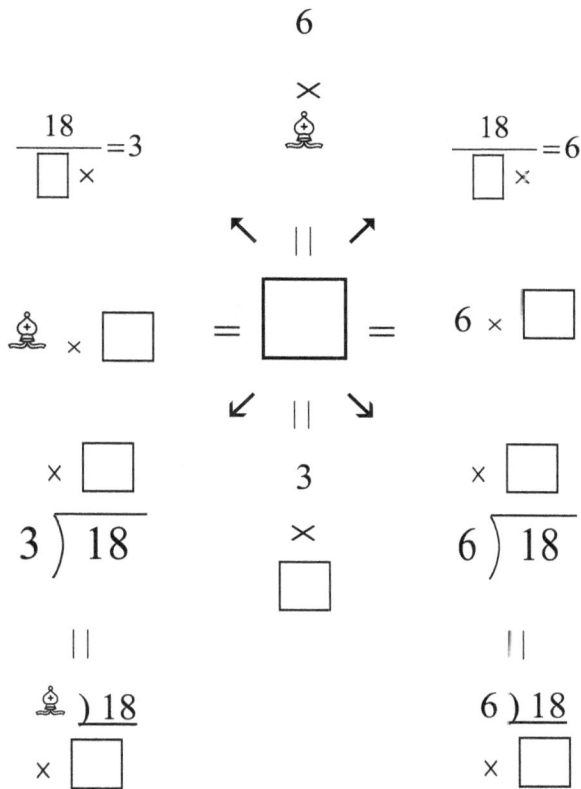

Ho Math Chess 何数棋谜 妈！我会棋谜式乘法啦！
Mom! I Learn Multiplication Using Math-Chess-Puzzles Connection!
Student's Name _____ Date _____
2007 - 2017 © Frank Ho, Amanda Ho, All rights reserved.   www.homathchess.com

Different ways of writing multiplication (Learning division while doing multiplications)

$$6$$
$$\times$$
$$4$$

$$\frac{24}{\boxed{}\times}=4 \qquad \frac{24}{\boxed{}\times}=6$$

$$4 \times \boxed{} = \boxed{} = \boxed{} \times 6$$

$$\times \boxed{} \qquad 4 \qquad \times \boxed{}$$
$$4\overline{)\,24} \qquad \times \qquad 6\overline{)\,24}$$
$$\boxed{}$$

$$||$$

$$4\,\underline{)\,24} \qquad\qquad 6\,\underline{)\,24}$$
$$\times \boxed{} \qquad\qquad \times \boxed{}$$

124

2007 - 2017 © Frank Ho, Amanda Ho, All rights reserved.   www.homathchess.com

Different ways of writing multiplication (Learning division while doing multiplications)

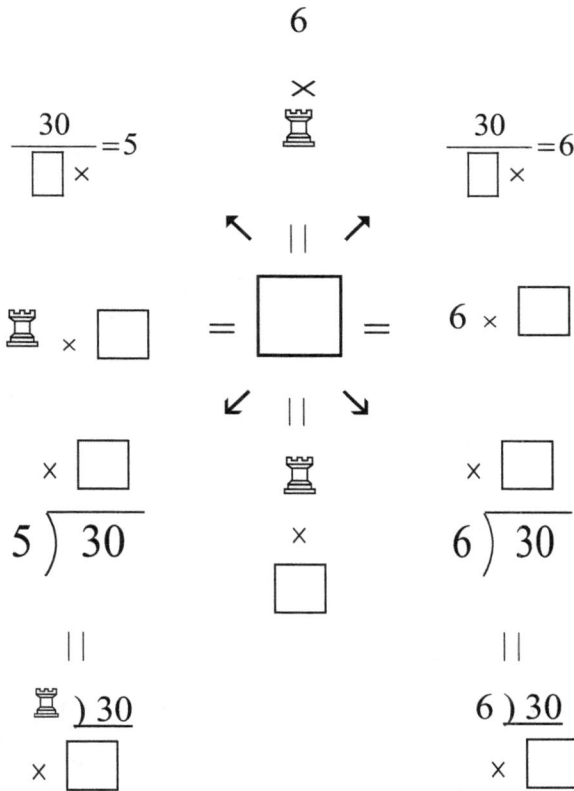

2007 - 2017 © Frank Ho, Amanda Ho, All rights reserved.　　www.homathchess.com

Different ways of writing multiplication (Learning division while doing multiplications)

$$6$$
$$\times$$
$$6$$

$$\frac{36}{\boxed{}\times}=6 \qquad\qquad \frac{36}{\boxed{}\times}=6$$

$$\nwarrow\ ||\ \nearrow$$

$$6\times\boxed{} \quad=\quad \boxed{} \quad=\quad 6\times\boxed{}$$

$$\swarrow\ ||\ \searrow$$

$$\times\boxed{} \qquad\qquad 6 \qquad\qquad \times\boxed{}$$

$$6\,\overline{)\,36} \qquad\qquad \times \qquad\qquad 6\,\overline{)\,36}$$

$$\boxed{}$$

$$|| \qquad\qquad\qquad\qquad ||$$

$$6\,\underline{)\,36} \qquad\qquad\qquad 6\,\underline{)\,36}$$
$$\times\boxed{} \qquad\qquad\qquad \times\boxed{}$$

Ho Math Chess 何数棋谜 妈！我会棋谜式乘法啦！
Mom! I Learn Multiplication Using Math-Chess-Puzzles Connection!

Student's Name _____ Date _____

2007 - 2017 © Frank Ho, Amanda Ho, All rights reserved.   www.homathchess.com

Different ways of writing multiplication (Learning division while doing multiplications)

$$\frac{42}{\square\times} = 6 \qquad \begin{array}{c}6\\ \times\\ 7\end{array} \qquad \frac{42}{\square\times} = 7$$

$$\nwarrow \; || \; \nearrow$$

$$6 \times \square \quad = \quad \boxed{} \quad = \quad 7 \times \square$$

$$\swarrow \; || \; \searrow$$

$$\begin{array}{c}\times \square\\ 6\,\overline{)\,42}\end{array} \qquad \begin{array}{c}7\\ \times\\ \square\end{array} \qquad \begin{array}{c}\times \square\\ 7\,\overline{)\,42}\end{array}$$

$$||$$

$$6\,\underline{)\,42} \qquad\qquad 7\,\underline{)\,42}$$
$$\times \square \qquad\qquad\quad \times \square$$

127

2007 - 2017 © Frank Ho, Amanda Ho, All rights reserved. www.homathchess.com

Different ways of writing multiplication (Learning division while doing multiplications)

$$6$$
$$\times$$
$$8$$

$$\frac{48}{\boxed{}\times}=6 \qquad\qquad \frac{48}{\boxed{}\times}=8$$

$$\nwarrow \quad || \quad \nearrow$$

$$6\times\boxed{} \qquad =\boxed{}= \qquad 8\times\boxed{}$$

$$\swarrow \quad || \quad \searrow$$

$$\times\boxed{} \qquad\qquad 8 \qquad\qquad \times\boxed{}$$

$$6\overline{)\,48} \qquad \begin{array}{c}\times\\ \boxed{}\end{array} \qquad 8\overline{)\,48}$$

$$|| \qquad\qquad\qquad\qquad ||$$

$$6\,\underline{)\,48} \qquad\qquad\qquad 8\,\underline{)\,48}$$
$$\times\boxed{} \qquad\qquad\qquad \times\boxed{}$$

Ho Math Chess　何数棋谜　妈！我会棋谜式乘法啦！
Mom! I Learn Multiplication Using Math-Chess-Puzzles Connection!

Student's Name _____ Date _____

2007 - 2017 © Frank Ho, Amanda Ho, All rights reserved.　　www.homathchess.com

Different ways of writing multiplication (Learning division while doing multiplications)

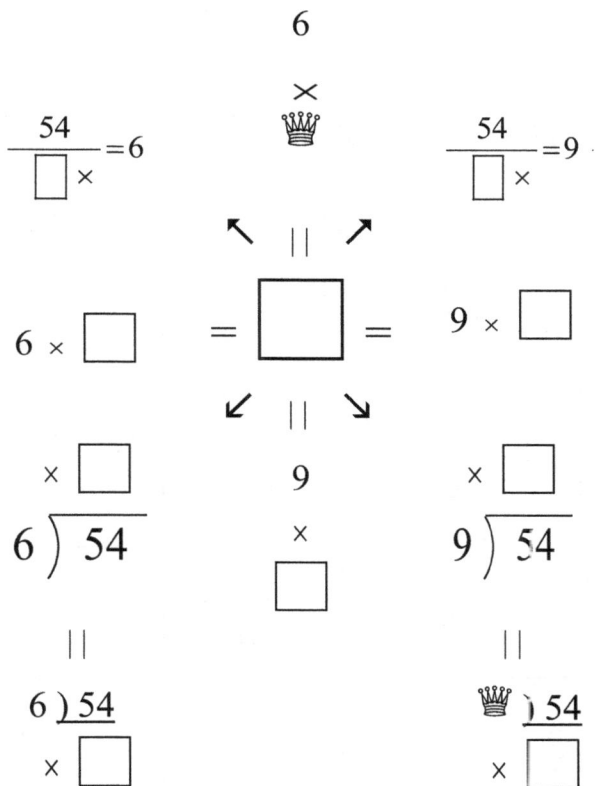

$$6$$
$$\times$$
♛

$$\frac{54}{\Box \times}=6 \qquad \Box = \qquad \frac{54}{\Box \times}=9$$

$$6 \times \Box \quad = \quad \boxed{} \quad = \quad 9 \times \Box$$

$$\times \Box \qquad 9 \qquad \times \Box$$
$$6\overline{)54} \qquad \times \qquad 9\overline{)54}$$
$$\Box$$

$$\Vert \qquad\qquad \Vert$$

$$6\,\overline{)54} \qquad\qquad ♛\,\overline{)54}$$
$$\times \Box \qquad\qquad \times \Box$$

Ho Math Chess   何数棋谜   妈！我会棋谜式乘法啦！
Mom! I Learn Multiplication Using Math-Chess-Puzzles Connection!

Student's Name _____ Date _____

2007 - 2017 © Frank Ho, Amanda Ho, All rights reserved.   www.homathchess.com

Counting 7's multiples

7, 14, 21, ☐, ☐, ☐, ☐, ☐, ☐

Fill in the following ☐ with a number.

Sequence	1	2	♞	4	♜	6	7	8	♛
Add 7	☐	14	☐	28	☐	42	☐	56	☐

Sequence	1	2	3	4	5	6	7	8	9
Add 7	7	☐	21	☐	35	☐	49	☐	63

Sequence	1	2	♝	4	♜	6	7	8	♛
Add 7	☐	14	☐	28	☐	42	☐	56	☐

Sequence	1	2	3	4	5	6	7	8	9
Add 7	7	☐	21	☐	35	☐	49	☐	63

Sequence	1	2	♝	4	♜	6	7	8	♛
Add 7	☐	14	☐	28	☐	42	☐	56	☐

Sequence	1	2	3	4	5	6	7	8	9
Add 7	7	☐	21	☐	35	☐	49	☐	63

7 times

$7 \times 1 = \square$	Seven times one is $\square$	$1 \times 7 = \square$	One times seven is $\square$
$7 \times 2 = \square$	Seven times two is $\square$	$2 \times 7 = \square$	Two times seven is $\square$
$7 \times 3 = \square$	Seven times three is $\square$	$3 \times 7 = \square$	Three times seven is $\square$
$7 \times 4 = \square$	Seven times four is $\square$	$4 \times 7 = \square$	Four times seven is $\square$
$7 \times 5 = \square$	Seven times five is $\square$	$5 \times 7 = \square$	Five times seven is $\square$
$7 \times 6 = \square$	Seven times six is $\square$	$6 \times 7 = \square$	Six times seven is $\square$
$7 \times 7 = \square$	Seven times seven is $\square$	$7 \times 7 = \square$	Seven times seven is $\square$
$7 \times 8 = \square$	Seven times eight is $\square$	$8 \times 7 = \square$	Eight times seven is $\square$
$7 \times 9 = \square$	Seven times nine is $\square$	$9 \times 7 = \square$	Nine times seven is $\square$

Ho Math Chess   何数棋谜   妈！我会棋谜式乘法啦！
Mom! I Learn Multiplication Using Math-Chess-Puzzles Connection!

Student's Name _____ Date _____

2007 - 2017 © Frank Ho, Amanda Ho, All rights reserved.   www.homathchess.com

7	1	7	2	7
X 1	X 7	X 2	X 7	X 3

6	7	7	7	9
X 7	X 5	X 7	X 8	X 7

6	8	7	4	♛
X 7	X 7	X 7	X 7	X 7

♜	6	7	8	9
X 7	X 7	X 3	X 7	X 7

7	7	♜	7	7
X 7	X 8	X 7	X 6	X 9

2007 - 2017 © Frank Ho, Amanda Ho, All rights reserved.　　www.homathchess.com

Oral practice

seven one seven	7 1 □	1 1 × 7 7 7 □ □
seven two fourteen	7 2 □	2 2 × 7 1 5 4 □ □ □
seven three twenty-one	7 ♘ □	3 3 × 7 2 3 1 □ □ □
seven four twenty-eight	7 4 □	4 4 × 7 3 0 8 □ □ □
seven five thirty-five	7 ♖ □	³5 5 × 7 3 8 5 □ □ □

2007 - 2017 © Frank Ho, Amanda Ho, All rights reserved.　www.homathchess.com

Oral practice

seven six forty-two	7 6 ☐	$\overset{4}{6}6$ $\times\ \ 7$ 462 ☐☐☐
seven seven forty-nine	7 7 ☐	7 7 $\times\ \ 7$ 539 ☐☐☐
seven eight fifty-six	7 8 ☐	8 8 $\times\ \ 7$ 616 ☐☐☐
seven nine sixty-three	7 ♛ ☐	9 9 $\times\ \ 7$ 693 ☐☐☐
seven five thirty-five	7 ♜ ☐	5 5 $\times\ \ 7$ 385 ☐☐☐
two six forty-two	7 6 ☐	6 6 $\times\ \ 7$ 462 ☐☐☐

2007 - 2017 © Frank Ho, Amanda Ho, All rights reserved.　　www.homathchess.com

Preparing for division

☐	☐	☐	☐	☐
X 2	X 5	X 4	X 6	X 3
14	35	28	42	21

☐	☐	☐	☐	☐
X 8	X 6	X 2	X ♕	X 7
56	42	14	63	49

☐	☐	☐	☐	☐
X 4	X 2	X ♖	X 2	X 7
28	14	35	14	49

☐	☐	☐	☐	☐
X 6	X 5	X 2	X 3	X 4
42	35	14	21	28

☐	☐	☐	☐	☐
X ♖	X 6	X ♗	X 4	X 7
35	42	21	28	49

2007 - 2017 © Frank Ho, Amanda Ho, All rights reserved.　www.homathchess.com

Fill in ☐ with answer.

Times	Grouping	Addition
$7 \times \$1 = \square$	$7 \text{ of } \square = 7$	$\$1 + \$1 + \$1 + \$1 + \$1 + \$1 + \$1 = \square$
$1 \times \$7 = \square$	$1 \text{ of } \square = 7$	$\$7$

Fill in ☐ with answer.

Expression	Grouping	Addition
$7 \times \$2$	$7 \text{ of } \square = 14$	$\$2 + \$2 + \$2 + \$2 + \$2 + \$2 + \$2 = \square$
$2 \times \$7$	$2 \text{ of } \square = 14$	$\$7 + \$7 = \square$

$7 \times 1 = \square = 1 \times 7 = \square$	$1 \times 7 = \square = 7 \times 1 = \square$
$7 \times \square = 14 = 2 \times \square = \square$	$2 \times \square = 14 = 7 \times \square = \square$

7 1 ☐	7 5 ☐	7 9 ☐	7 4 ☐	7 8 ☐
7 2 ☐	7 6 ☐	7 1 ☐	7 ♖ ☐	7 ♕ ☐
7 ♗ ☐	7 7 ☐	7 2 ☐	7 6 ☐	7 1 ☐
7 4 ☐	7 8 ☐	7 3 ☐	7 7 ☐	7 2 ☐

Ho Math Chess　何数棋谜　妈！我会棋谜式乘法啦！

Mom! I Learn Multiplication Using Math-Chess-Puzzles Connection!

Student's Name _____ Date _____

2007 - 2017 © Frank Ho, Amanda Ho, All rights reserved.　www.homathchess.com

Fill in ☐ with answer.

Times	Grouping	Addition
$7 \times \$3 = $ ☐	7 of ☐ $= 21$	$\$3 + \$3 + \$3 + \$3 + \$3 + \$3 + \$3 = $ ☐
$3 \times \$7 = $ ☐	3 of ☐ $= 21$	$\$7 + \$7 + \$7 = $ ☐

Fill in ☐ with answer.

Expression	Grouping	Addition
$7 \times \$4$	7 of ☐ $= 28$	$\$4 + \$4 + \$4 + \$4 + \$4 + \$4 + \$4 = $ ☐
$4 \times \$7$	4 of ☐ $= 28$	$\$7 + \$7 + \$7 + \$7 = $ ☐

$7 \times ♘ = $ ☐ $= 3 \times 7 = $ ☐	$3 \times 7 = $ ☐ $= 7 \times 3 = $ ☐
$7 \times $ ☐ $= 28 = 4 \times $ ☐ $= $ ☐	$4 \times $ ☐ $= 28 = 7 \times $ ☐ $= $ ☐

7 1 ☐	7 5 ☐	7 ♛ ☐	7 4 ☐	7 8 ☐
7 2 ☐	7 6 ☐	7 1 ☐	7 ♖ ☐	7 ♛ ☐
7 ♗ ☐	7 7 ☐	7 2 ☐	7 6 ☐	7 1 ☐
7 4 ☐	7 8 ☐	7 3 ☐	7 7 ☐	7 2 ☐

2007 - 2017 © Frank Ho, Amanda Ho, All rights reserved.   www.homathchess.com

Fill in ☐ with answer.

Times	Grouping	Addition
$7 \times \$5 = \square$	7 of $\square$ = 35	$\$5 + \$5 + \$5 + \$5 + \$5 + \$5 + \$5 = \square$
$5 \times \$7 = \square$	5 of $\square$ = 35	$\$7 + \$7 + \$7 + \$7 + \$7 = \square$

Times	Grouping	Addition
$7 \times \$6 = \square$	6 of $\square$ = 42	$\$6 + \$6 + \$6 + \$6 + \$6 + \$6 + \$6 = \square$
$6 \times \$7 = \square$	6 of $\square$ = 42	$\$6 + \$6 + \$6 + \$6 + \$6 + \$6 + \$6 = \square$

$7 \times ♖ = \square = ♖ \times 7 = \square$	$5 \times 7 = \square = 7 \times 5 = \square$
$7 \times \square = 42 = 6 \times \square = \square$	$6 \times \square = 42 = 7 \times \square = \square$

7 ♙ $\square$	7 5 $\square$	7 9 $\square$	7 4 $\square$	7 8 $\square$
7 2 $\square$	7 6 $\square$	7 1 $\square$	7 ♖ $\square$	7 ♛ $\square$
7 3 $\square$	7 7 $\square$	7 2 $\square$	7 6 $\square$	7 ♙ $\square$
7 4 $\square$	7 8 $\square$	7 3 $\square$	7 7 $\square$	7 2 $\square$

Student's Name _____ Date _____

2007 - 2017 © Frank Ho, Amanda Ho, All rights reserved.   www.homathchess.com

Fill in _____ and ☐ with answers.

Times	Grouping	Addition
$7 \times \$7 = \square$	7 of $\square$ = 49	$\$7 + \$7 + \$7 + \$7 + \$7 + \$7 + \$7 = \square$
$7 \times \$7 = \square$	7 of $\square$ = 49	$\$7 + \$7 + \$7 + \$7 + \$7 + \$7 + \$7 = \square$

Times	Grouping	Addition
$7 \times \$8 = \square$	7 of $\square$ = 56	$\$8 + \$8 + \$8 + \$8 + \$8 + \$8 + \$8 = \square$
$8 \times \$7 = \square$	8 of $\square$ = 56	$\$7 + \$7 + \$7 + \$7 + \$7 + \$7 + \$7 + \$7 = \square$

$7 \times 7 = \square = 7 \times 7 = \square$	$7 \times 7 = \square = 7 \times 7 = \square$
$7 \times \square = 56 = 8 \times \square = \square$	$8 \times \square = 56 = 7 \times \square = \square$

7 ♙ ☐	7 5 ☐	7 9 ☐	7 4 ☐	7 8 ☐
7 2 ☐	7 6 ☐	7 ♙ ☐	7 ♖ ☐	7 ♛ ☐
7 3 ☐	7 7 ☐	7 2 ☐	7 6 ☐	7 1 ☐
7 4 ☐	7 8 ☐	7 ♞ ☐	7 7 ☐	7 2 ☐

2007 - 2017 © Frank Ho, Amanda Ho, All rights reserved.   www.homathchess.com

Fill in ☐ with answer.

Times	Grouping	Addition
$7 \times \$9 = \square$	7 of $\square$ = 63	$\$9 + \$9 + \$9 + \$9 + \$9 + \$9 + \$9 = \square$
$9 \times \$7 = \square$	9 of $\square$ = 63	$\$7 + \$7 + \$7 + \$7 + \$7 + \$7 + \$7 + \$7 + \$7 = \square$

Times	Grouping	Addition
$7 \times \$8 = \square$	7 of $\square$ = 56	$\$8 + \$8 + \$8 + \$8 + \$8 + \$8 + \$8 = \square$
$8 \times \$7 = \square$	8 of $\square$ = 56	$\$7 + \$7 + \$7 + \$7 + \$7 + \$7 + \$7 + \$7 = \square$

$7 \times 9 = \square = ♛ \times 7 = \square$	$♛ \times 7 = \square = 7 \times 9 = \square$
$9 \times \square = 63 = 9 \times \square = \square$	$7 \times \square = 63 = 9 \times \square = \square$

7 ♙ $\square$	7 5 $\square$	7 9 $\square$	7 4 $\square$	7 8 $\square$
7 2 $\square$	7 6 $\square$	7 1 $\square$	7 ♜ $\square$	7 ♛ $\square$
7 ♝ $\square$	7 7 $\square$	7 2 $\square$	7 6 $\square$	7 ♙ $\square$
7 4 $\square$	7 8 $\square$	7 ♝ $\square$	7 7 $\square$	7 2 $\square$

2007 - 2017 © Frank Ho, Amanda Ho, All rights reserved.   www.homathchess.com

Preparing for division

□ X 7 = 7	X □ 7)7	□)7 ___ X 7
□ X 7 = 14	X □ 7)14	□)14 ___ X 7
□ X 7 = 21	X □ 7)21	□)21 ___ X 7
□ X 7 = 28	X □ 7)28	□)28 ___ X 7
□ X 7 = 35	X □ 7)35	□)35 ___ X 7
□ X 7 = 42	X □ 7)42	□)42 ___ X 7
□ X 7 = 49	X □ 7)49	□)49 ___ X 7

2007 - 2017 © Frank Ho, Amanda Ho, All rights reserved.   www.homathchess.com

Preparing for division

☐ X 7 = 56	X ☐ 7)56	☐) 56 X 7
☐ X 7 = 63	X ☐ 7)63	☐) 63 X 7
☐ X 7 = 7	X ☐ 7)7	☐) 7 X 7
☐ X 7 = 14	X ☐ 7)14	☐) 14 X 7
☐ X 7 = 21	X ☐ 7)21	☐) 21 X 7
☐ X 7 = 28	X ☐ 7)28	☐) 28 X 7
☐ X 7 = 35	X ☐ 7)35	☐) 35 X 7

Ho Math Chess 何数棋谜 妈！我会棋谜式乘法啦！

Mom! I Learn Multiplication Using Math-Chess-Puzzles Connection!

Student's Name _____ Date _____

2007 - 2017 © Frank Ho, Amanda Ho, All rights reserved. www.homathchess.com

Cross multiplication

12　　　12 ↖ ↗ $\frac{6}{2} = \frac{6}{2}$	□　14 ↖ ↗ $\frac{7}{7} = \frac{2}{2}$	□　21 ↖ ↗ $\frac{7}{7} = \frac{3}{3}$	□　28 ↖ ↗ $\frac{7}{7} = \frac{4}{4}$
□　35 ↖ ↗ $\frac{7}{7} = \frac{5}{5}$	□　42 ↖ ↗ $\frac{7}{7} = \frac{6}{6}$	□　49 ↖ ↗ $\frac{7}{7} = \frac{7}{7}$	□　63 ↖ ↗ $\frac{7}{7} = \frac{9}{9}$
□　35 ↖ ↗ $\frac{7}{7} = \frac{5}{5}$	□　28 ↖ ↗ $\frac{7}{7} = \frac{4}{4}$	□　49 ↖ ↗ $\frac{7}{7} = \frac{7}{7}$	□　56 ↖ ↗ $\frac{7}{7} = \frac{8}{8}$
□　42 ↖ ↗ $\frac{7}{7} = \frac{6}{6}$	□　56 ↖ ↗ $\frac{7}{7} = \frac{8}{8}$	□　49 ↖ ↗ $\frac{7}{7} = \frac{9}{9}$	□　21 ↖ ↗ $\frac{7}{7} = \frac{3}{3}$

Different ways of writing multiplication (Learning division while doing multiplications)

$$7 \times 2$$

$$\frac{14}{\boxed{} \times} = 2 \qquad\qquad \frac{14}{\boxed{} \times} = 2$$

$$2 \times \boxed{} \quad = \quad \boxed{} \quad = \quad 7 \times \boxed{}$$

$$\times \boxed{} \qquad\qquad \times \boxed{}$$

$$2\overline{)14} \qquad\qquad 2 \times \boxed{} \qquad\qquad 7\overline{)14}$$

$$\|\qquad\qquad\qquad\qquad\qquad\qquad\|$$

$$2\,)\,14 \qquad\qquad\qquad 7\,)\,14$$

$$\times \boxed{} \qquad\qquad\qquad \times \boxed{}$$

Different ways of writing multiplication (Learning division while doing multiplications)

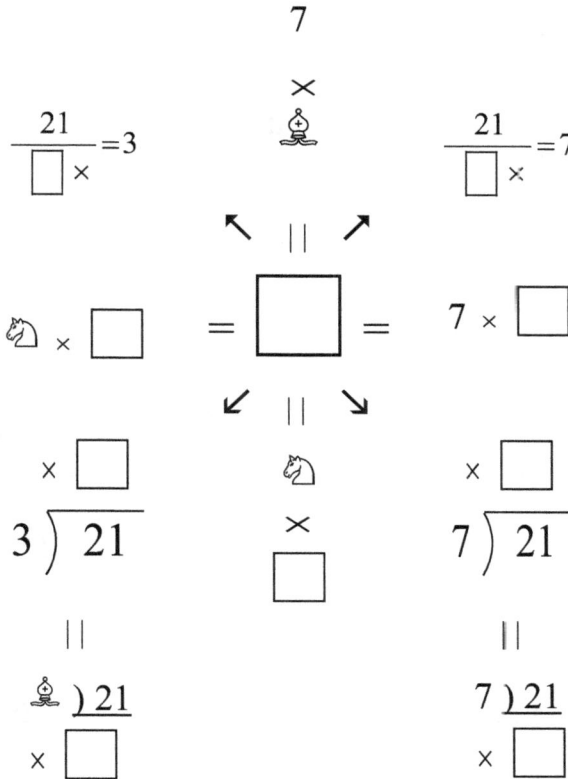

2007 - 2017 © Frank Ho, Amanda Ho, All rights reserved.   www.homathchess.com

Different ways of writing multiplication (Learning division while doing multiplications)

$$\frac{28}{\boxed{}\times} = 4 \qquad \begin{array}{c} 7 \\ \times \\ 4 \end{array} \qquad \frac{28}{\boxed{}\times} = 7$$

$$4 \times \boxed{} \quad = \boxed{} = \quad \boxed{} \times 7$$

$$\begin{array}{c} \times \boxed{} \\ 4\overline{)\,28} \end{array} \qquad \begin{array}{c} 4 \\ \times \\ \boxed{} \end{array} \qquad \begin{array}{c} \times \boxed{} \\ 7\overline{)\,28} \end{array}$$

$$\begin{array}{c} || \\ 4\,\underline{)\,28} \\ \times \boxed{} \end{array} \qquad\qquad \begin{array}{c} || \\ 7\,\underline{)\,28} \\ \times \boxed{} \end{array}$$

2007 - 2017 © Frank Ho, Amanda Ho, All rights reserved.   www.homathchess.com

Different ways of writing multiplication (Learning division while doing multiplications)

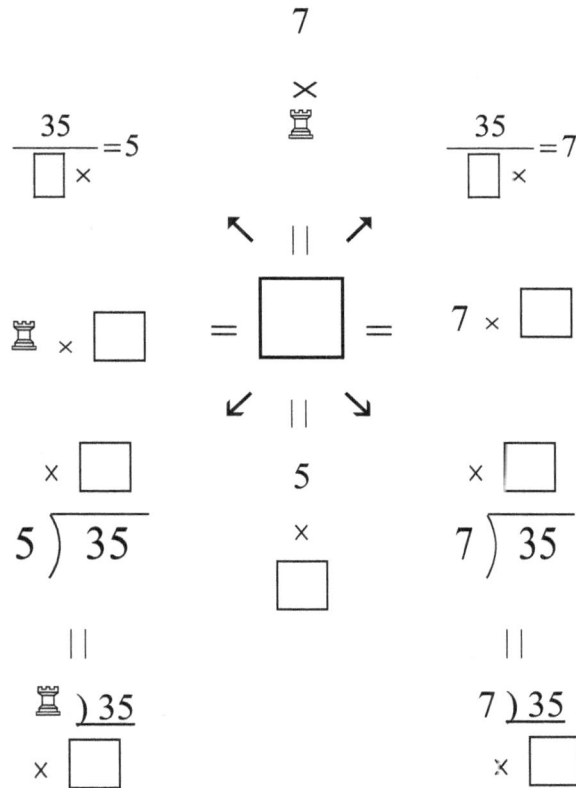

$$\frac{35}{\square \times} = 5 \qquad \qquad 7 \times \text{♜} \qquad \qquad \frac{35}{\square \times} = 7$$

$$\text{♜} \times \square \quad = \quad \boxed{} \quad = \quad 7 \times \square$$

$$\times \square \qquad \qquad 5 \qquad \qquad \times \square$$

$$5 \overline{)\,35} \qquad \qquad \times \qquad \qquad 7 \overline{)\,35}$$

$$\square$$

$$\text{♜} \,)\,35 \qquad \qquad \qquad 7\,)\,35$$
$$\times \square \qquad \qquad \qquad \times \square$$

2007 - 2017 © Frank Ho, Amanda Ho, All rights reserved.   www.homathchess.com

Different ways of writing multiplication (Learning division while doing multiplications)

$$\frac{42}{\square \times} = 6 \qquad \begin{array}{c} 7 \\ \times \\ 6 \end{array} \qquad \frac{42}{\square \times} = 7$$

$$6 \times \square \quad = \quad \boxed{} \quad = \quad 7 \times \square$$

$$\begin{array}{c} \times \square \\ 6 \overline{)\,42} \end{array} \qquad \begin{array}{c} 6 \\ \times \\ \square \end{array} \qquad \begin{array}{c} \times \square \\ 7 \overline{)\,42} \end{array}$$

$$6\,)\,42 \qquad\qquad 7\,)\,42$$
$$\times \square \qquad\qquad\qquad \times \square$$

2007 - 2017 © Frank Ho, Amanda Ho, All rights reserved.　　www.homathchess.com

Different ways of writing multiplication (Learning division while doing multiplications)

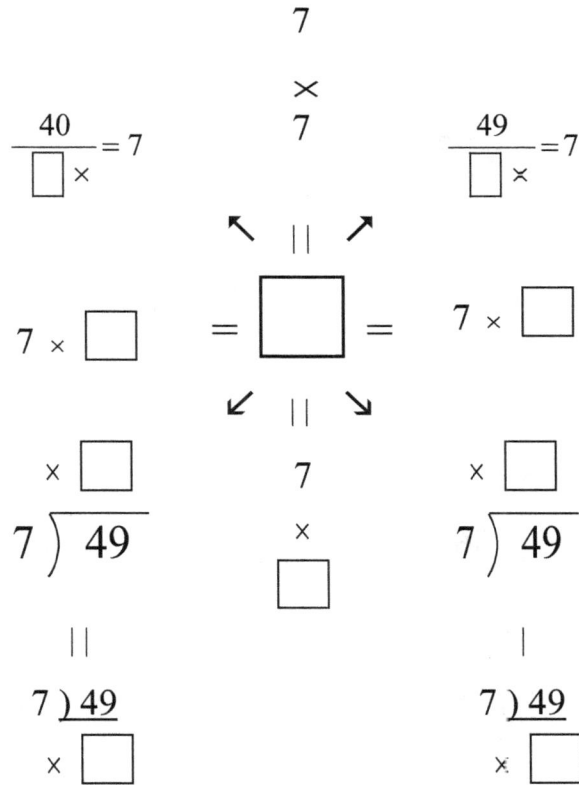

$$7 \times 7$$

$$\frac{40}{\square \times} = 7 \qquad \frac{49}{\square \times} = 7$$

$$7 \times \square \quad = \boxed{} = \quad 7 \times \square$$

$$\times \square \qquad 7 \qquad \times \square$$

$$7\overline{)49} \qquad \times \qquad 7\overline{)49}$$
$$\square$$

$$7\,)\,49 \qquad\qquad 7\,)\,49$$
$$\times \square \qquad\qquad \times \square$$

2007 - 2017 © Frank Ho, Amanda Ho, All rights reserved.　　www.homathchess.com

Different ways of writing multiplication (Learning division while doing multiplications)

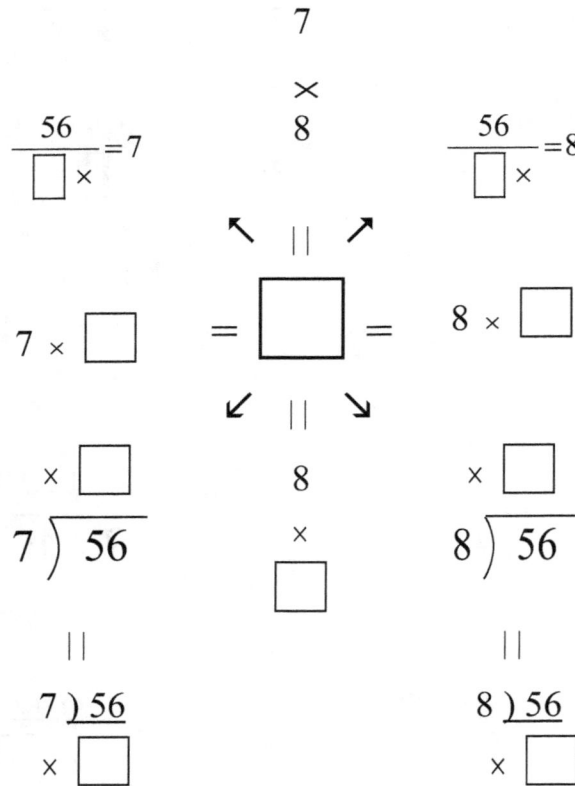

Ho Math Chess   何数棋谜   妈！我会棋谜式乘法啦！
Mom! I Learn Multiplication Using Math-Chess-Puzzles Connection!

Student's Name _____ Date _____

2007 - 2017 © Frank Ho, Amanda Ho, All rights reserved.   www.homathchess.com

Different ways of writing multiplication (Learning division while doing multiplications)

$$7 \times 9$$

$$\frac{63}{\square \times} = 7 \qquad \qquad \frac{63}{\square \times} = 9$$

$$7 \times \square \quad = \quad \boxed{} \quad = \quad 9 \times \square$$

$$\times \square \qquad \qquad \times \square$$

$$7 \overline{)\, 63} \qquad 9 \times \square \qquad 9 \overline{)\, 63}$$

$$7 \underline{)\, 63} \qquad \qquad \qquad 9 \underline{)\, 63}$$

$$\times \square \qquad \qquad \qquad \times \square$$

2007 - 2017 © Frank Ho, Amanda Ho, All rights reserved.   www.homathchess.com

Counting 8's multiples

8, 16, 24, ☐,☐,☐,☐,☐,☐

Fill in the following ☐ with a number.

Sequence	♙	2	♗	4	5	6	7	8	♛
Add 8	☐	16	☐	32	☐	48	☐	64	☐

Sequence	1	2	♞	4	♖	6	7	8	9
Add 8	8	☐	24	☐	40	☐	56	☐	72

Sequence	♙	2	3	4	5	6	7	8	♛
Add 8	☐	16	☐	32	☐	48	☐	64	☐

Sequence	1	2	♗	4	♖	6	7	8	9
Add 8	8	☐	24	☐	40	☐	56	☐	72

Sequence	♙	2	3	4	5	6	7	8	♛
Add 8	☐	16	☐	32	☐	48	☐	64	☐

Sequence	1	2	♗	4	5	6	7	8	9
Add 8	8	☐	24	☐	40	☐	56	☐	72

Ho Math Chess 何数棋谜 妈！我会棋谜式乘法啦！

Mom! I Learn Multiplication Using Math-Chess-Puzzles Connection!

Student's Name _____ Date _____

2007 - 2017 © Frank Ho, Amanda Ho, All rights reserved.   www.homathchess.com

8 times

8 × 1 = ☐	Eight times one is ☐	1 × 8 = ☐	One times eight is ☐
8 × 2 = ☐	Eight times two is ☐	2 × 8 = ☐	Two times eight is ☐
8 × 3 = ☐	Eight times three is ☐	♟ × 8 = ☐	Three times eight is ☐
8 × 4 = ☐	Eight times four is ☐	4 × 8 = ☐	Four times eight is ☐
8 × ♖ = ☐	Eight times five is ☐	5 × 8 = ☐	Five times eight is ☐
8 × 6 = ☐	Eight times six is ☐	6 × 8 = ☐	Six times eight is ☐
8 × 7 = ☐	Eight times seven is ☐	7 × 8 = ☐	Seven times eight is ☐
8 × 8 = ☐	Eight times eight is ☐	8 × 8 = ☐	Eight times eight is ☐
8 × ♛ = ☐	Eight times nine is ☐	9 × 8 = ☐	Nine times eight is ☐

$$8 \times 1 \quad 1 \times 8 \quad 2 \times 8 \quad 8 \times 2 \quad ♞ \times 8$$

$$5 \times 8 \quad 8 \times 6 \quad 6 \times 8 \quad 8 \times 8 \quad ♖ \times 8$$

$$4 \times 8 \quad 8 \times ♛ \quad 5 \times 8 \quad 8 \times 6 \quad 6 \times 8$$

153

Ho Math Chess　何数棋谜　妈！我会棋谜式乘法啦！

Mom! I Learn Multiplication Using Math-Chess-Puzzles Connection!

Student's Name _____ Date _____

2007 - 2017 © Frank Ho, Amanda Ho, All rights reserved.　www.homathchess.com

8	1	8	2	8
X 1	X 8	X 2	X 8	X 3

8	6	7	8	♛
X 4	X 8	X 8	X 8	X 8

8	8	8	8	8
X ♜	X 8	X 7	X 8	X 6

5	8	7	4	9
X 8	X 6	X 8	X 8	X 8

7	♜	8	8	8
X 8	X 8	X 6	X 6	X 9

2007 - 2017 © Frank Ho, Amanda Ho, All rights reserved.　www.homathchess.com

Oral practice

eight one eight	8 1 ☐	1 1 × 8 88☐☐
eight two sixteen	8 2 ☐	2 2 × 8 176☐☐☐
eight three twenty-four	8 ♞ ☐	3 3 × 8 264☐☐☐
eight four thirty-two	8 4 ☐	4 4 × 8 352☐☐☐
eight five forty	8 5 ☐	⁴ 5 5 × 8 440☐☐☐

Ho Math Chess 何数棋谜 妈！我会棋谜式乘法啦！
Mom! I Learn Multiplication Using Math-Chess-Puzzles Connection!

Student's Name _____ Date _____

2007 - 2017 © Frank Ho, Amanda Ho, All rights reserved.   www.homathchess.com

Oral practice

eight six forty-eight	8 6 ☐	4 6 6 × 8 528☐☐☐
eight seven fifty-six	8 7 ☐	7 7 × 8 616☐☐☐
eight eight sixty-four	8 8 ☐	8 8 × 8 704☐☐☐
eight nine seventy-two	8 ♛ ☐	9 9 × 8 792☐☐☐
eight two sixteen	8 2 ☐	2 2 × 8 176☐☐☐
eight three twenty-four	8 ♞ ☐	3 3 × 8 264☐☐☐

Ho Math Chess   何数棋谜   妈！我会棋谜式乘法啦！
Mom! I Learn Multiplication Using Math-Chess-Puzzles Connection!

Student's Name _____ Date _____

2007 - 2017 © Frank Ho, Amanda Ho, All rights reserved.   www.homathchess.com

Fill in ☐ with answer.

Times	Grouping	Addition
$8 \times \$1 = \Box$	$8 \text{ of } \Box = 8$	$\$1 + \$1 + \$1 + \$1 + \$1 + \$1 + \$1 + \$1 = \Box$
$1 \times \$8 = \Box$	$1 \text{ of } \Box = 8$	$\$8$

Fill in ☐ with answer.

Expression	Grouping	Addition
$8 \times \$2$	$8 \text{ of } \Box = 16$	$\$2 + \$2 + \$2 + \$2 + \$2 + \$2 + \$2 + \$2 = \Box$
$2 \times \$8$	$2 \text{ of } \Box = 16$	$\$8 + \$8 = \Box$

$8 \times 1 = \Box = 1 \times 8 = \Box$	$1 \times 8 = \Box = 8 \times 1 = \Box$
$8 \times \Box = 16 = 2 \times \Box = \Box$	$2 \times \Box = 16 = 8 \times \Box = \Box$

8 ♙ ☐	8 ♖ ☐	8 9 ☐	8 4 ☐	8 8 ☐
8 2 ☐	8 6 ☐	8 ♙ ☐	8 ♖ ☐	8 ♕ ☐
8 ♘ ☐	8 7 ☐	8 2 ☐	8 6 ☐	8 ♙ ☐
8 4 ☐	8 8 ☐	8 3 ☐	8 7 ☐	8 2 ☐

2007 - 2017 © Frank Ho, Amanda Ho, All rights reserved.   www.homathchess.com

Fill in ☐ with answer.

Times	Grouping	Addition
8 × $3 = ☐	8 of ☐ = 24	$3 + $3 + $3 + $3 + $3 + $3 + $3 + $3 = ☐
♘ × $8 = ☐	3 of ☐ = 24	$8 + $8 + $8 = ☐

Fill in ☐ with answer.

Times	Grouping	Addition
8 × $4	8 of ☐ = 32	$4 + $4 + $4 + $4 + $4 + $4 + $4 + $4 = ☐
4 × $8	4 of ☐ = 32	$8 + $8 + $8 + $8 = ☐

8 × 3 = ☐ = ♘ × 8 = ☐	3 × 8 = ☐ = 8 × 3 = ☐
8 × ☐ = 32 = 4 × ☐ = ☐	4 × ☐ = 32 = 8 × ☐ = ☐

8 1 ☐	8 ♖ ☐	8 9 ☐	8 4 ☐	8 8 ☐
8 2 ☐	8 6 ☐	8 1 ☐	8 5 ☐	8 9 ☐
8 ♞ ☐	8 7 ☐	8 2 ☐	8 6 ☐	8 ♙ ☐
8 4 ☐	8 8 ☐	8 ♞ ☐	8 7 ☐	8 2 ☐

2007 - 2017 © Frank Ho, Amanda Ho, All rights reserved.　　www.homathchess.com

Fill in ☐ with answer.

Times	Grouping	Addition
$8 \times \$5 =$ ☐	8 of ☐ $= 40$	$\$5 + \$5 + \$5 + \$5 + \$5 + \$5 + \$5 + \$5 =$ ☐
$5 \times \$8 =$ ☐	5 of ☐ $= 40$	$\$8 + \$8 + \$8 + \$8 + \$8 =$ ☐

Times	Grouping	Addition
$8 \times \$6 =$ ☐	8 of ☐ $= 48$	$\$6 + \$6 + \$6 + \$5 + \$6 + \$6 + \$6 + \$6 =$ ☐
$6 \times \$8 =$ ☐	6 of ☐ $= 48$	$\$8 + \$8 + \$8 + \$8 + \$8 + \$8 =$ ☐

$8 \times$ ♖ $=$ ☐ $= 5 \times 8 =$ ☐	$5 \times 8 =$ ☐ $= 8 \times 5 =$ ☐
$8 \times$ ☐ $= 48 = 6 \times$ ☐ $=$ ☐	$6 \times$ ☐ $= 48 = 8 \times$ ☐ $=$ ☐

8 1 ☐	8 ♖ ☐	8 9 ☐	8 4 ☐	8 8 ☐
8 2 ☐	8 6 ☐	8 ♙ ☐	8 5 ☐	8 ♕ ☐
8 ♗ ☐	8 7 ☐	8 2 ☐	8 6 ☐	8 1 ☐
8 4 ☐	8 8 ☐	8 3 ☐	8 7 ☐	8 2 ☐

Ho Math Chess   何数棋谜   妈！我会棋谜式乘法啦！
Mom! I Learn Multiplication Using Math-Chess-Puzzles Connection!

Student's Name _____ Date _____

2007 - 2017 © Frank Ho, Amanda Ho, All rights reserved.   www.homathchess.com

Fill in _____ and ☐ with answers.

Times	Grouping	Addition
8 × $7 = ☐	8 of ☐ = 56	$7 + $7 + $7 + $7 + $7 + $7 + $7 + $7 = ☐
7 × $8 = ☐	7 of ☐ = 56	$8 + $8 + $8 + $8 + $8 + $8 + $8 = ☐

Times	Grouping	Addition
8 × $8 = ☐	8 of ☐ = 64	$8 + $8 + $8 + $8 + $8 + $8 + $8 + $8 = ☐
8 × $8 = ☐	8 of ☐ = 64	$8 + $8 + $8 + $8 + $8 + $8 + $8 + $8 = ☐

8 × 7 = ☐ = 7 × 8 = ☐	7 × 8 = ☐ = 8 × 7 = ☐
8 × ☐ = 64 = 8 × ☐ = ☐	8 × ☐ = 64 = 8 × ☐ = ☐

8 1 ☐	8 5 ☐	8 ♛ ☐	8 4 ☐	8 8 ☐
8 2 ☐	8 6 ☐	8 1 ☐	8 ♜ ☐	8 9 ☐
8 ♞ ☐	8 7 ☐	8 2 ☐	8 6 ☐	8 ♙ ☐
8 4 ☐	8 8 ☐	8 3 ☐	8 7 ☐	8 2 ☐

2007 - 2017 © Frank Ho, Amanda Ho, All rights reserved.   www.homathchess.com

Fill in ☐ with answer.

Times	Grouping	Addition
$8 \times \$♛ = \square$	8 of $\square$ = 72	$\$9 + \$9 + \$9 + \$9 + \$9 + \$9 + \$9 + \$9 = \square$
$9 \times \$8 = \square$	9 of $\square$ = 72	$\$8 + \$8 + \$8 + \$8 + \$8 + \$8 + \$8 + \$8 + \$8 = \square$

Times	Grouping	Addition
$8 \times \$8 = \square$	8 of $\square$ = 64	$\$8 + \$8 + \$8 + \$8 + \$8 + \$8 + \$8 + \$8 = \square$
$8 \times \$8 = \square$	8 of $\square$ = 64	$\$8 + \$8 + \$8 + \$8 + \$8 + \$8 + \$8 + \$8 = \square$

$8 \times 9 = \square = 9 \times 8 = \square$	$9 \times 8 = \square = 8 \times 9 = \square$
$8 \times \square = 64 = 8 \times \square = \square$	$8 \times \square = 64 = 8 \times \square = \square$

8 ♟ ☐	8 ♜ ☐	8 ♛ ☐	8 4 ☐	8 8 ☐
8 2 ☐	8 6 ☐	8 1 ☐	8 ♜ ☐	8 9 ☐
8 ♝ ☐	8 7 ☐	8 2 ☐	8 6 ☐	8 ♟ ☐
8 4 ☐	8 8 ☐	8 3 ☐	8 7 ☐	8 2 ☐

Ho Math Chess 何数棋谜 妈！我会棋谜式乘法啦！
Mom! I Learn Multiplication Using Math-Chess-Puzzles Connection!

Student's Name _____ Date _____

2007 - 2017 © Frank Ho, Amanda Ho, All rights reserved. www.homathchess.com

Preparing for division

☐	☐	☐	☐	☐
X 5	X 2	X 4	X 9	X ♝
40	16	32	72	24

☐	☐	☐	☐	☐
X 7	X 6	X 8	X ♝	X 8
56	48	24	6	48

☐	☐	☐	☐	☐
X 4	X 8	X 5	X 2	X 7
32	72	40	16	56

☐	☐	☐	☐	☐
X 8	X 8	X 8	X 3	X 8
56	48	32	24	16

☐	☐	☐	☐	☐
X 4	X ♜	X 4	X 2	X 8
32	40	32	16	56

Ho Math Chess 　何数棋谜　妈！我会棋谜式乘法啦！
Mom! I Learn Multiplication Using Math-Chess-Puzzles Connection!
Student's Name _____ Date _____

2007 - 2017 © Frank Ho, Amanda Ho, All rights reserved.　www.homathchess.com

Preparing for division

□ X 8= 8	X □ 8)8	□)8 X 8
□ X 8 = 16	X □ 8)16	□)16 X 8
□ X 8 = 24	X □ 8)24	□)24 X 8
□ X 8 = 32	X □ 8)32	□)32 X 8
□ X 8 = 40	X □ 8)40	□)40 X 8
□ X 8 = 48	X □ 8)48	□)48 X 8
□ X 8 = 56	X □ 8)56	□)56 X 8

2007 - 2017 © Frank Ho, Amanda Ho, All rights reserved.   www.homathchess.com

Preparing for division

☐ X 8 = 64	X ☐ 8 ⟌ 64	☐) 64 X 8
☐ X 8 = 72	X ☐ 8 ⟌ 72	☐) 72 X 8
☐ X 8 = 16	X ☐ 8 ⟌ 16	☐) 16 X 8
☐ X 8 = 24	X ☐ 8 ⟌ 24	☐) 24 X 8
☐ X 8 = 32	X ☐ 8 ⟌ 32	☐) 32 X 8
☐ X 8 = 40	X ☐ 8 ⟌ 40	☐) 40 X 8
☐ X 8 = 48	X ☐ 8 ⟌ 48	☐) 48 X 8

Ho Math Chess　何数棋谜　妈！我会棋谜式乘法啦！
Mom! I Learn Multiplication Using Math-Chess-Puzzles Connection!

Student's Name _____ Date _____

2007 - 2017 © Frank Ho, Amanda Ho, All rights reserved.　www.homathchess.com

Cross multiplication

12 ↖　↗ 12 $\frac{6}{2} = \frac{6}{2}$	☐　16 ↖　↗ $\frac{8}{8} = \frac{2}{2}$	☐　24 ↖　↗ $\frac{8}{8} = \frac{3}{3}$	☐　32 ↖　↗ $\frac{8}{8} = \frac{4}{4}$
☐　40 ↖　↗ $\frac{8}{8} = \frac{5}{5}$	☐　48 ↖　↗ $\frac{8}{8} = \frac{6}{6}$	☐　56 ↖　↗ $\frac{8}{8} = \frac{7}{7}$	☐　72 ↖　↗ $\frac{8}{8} = \frac{9}{9}$
☐　40 ↖　↗ $\frac{8}{8} = \frac{5}{5}$	☐　32 ↖　↗ $\frac{8}{8} = \frac{4}{4}$	☐　56 ↖　↗ $\frac{8}{8} = \frac{7}{7}$	☐　64 ↖　↗ $\frac{8}{8} = \frac{8}{8}$
☐　48 ↖　↗ $\frac{8}{8} = \frac{6}{6}$	☐　72 ↖　↗ $\frac{8}{8} = \frac{9}{9}$	☐　64 ↖　↗ $\frac{8}{8} = \frac{8}{8}$	☐　24 ↖　↗ $\frac{8}{8} = \frac{3}{3}$

Different ways of writing multiplication (Learning division while doing multiplications)

$$8 \times 2$$

$$\frac{16}{\square \times} = 2 \qquad \frac{16}{\square \times} = 8$$

$$2 \times \square \ = \ \boxed{} \ = \ 8 \times \square$$

$$\times \square \qquad\qquad \times \square$$

$$2\overline{)16} \qquad 2 \times \square \qquad 8\overline{)16}$$

$$2\,)\,16 \qquad\qquad 8\,)\,16$$
$$\times \square \qquad\qquad\qquad \times \square$$

2007 - 2017 © Frank Ho, Amanda Ho, All rights reserved.   www.homathchess.com

Different ways of writing multiplication (Learning division while doing multiplications)

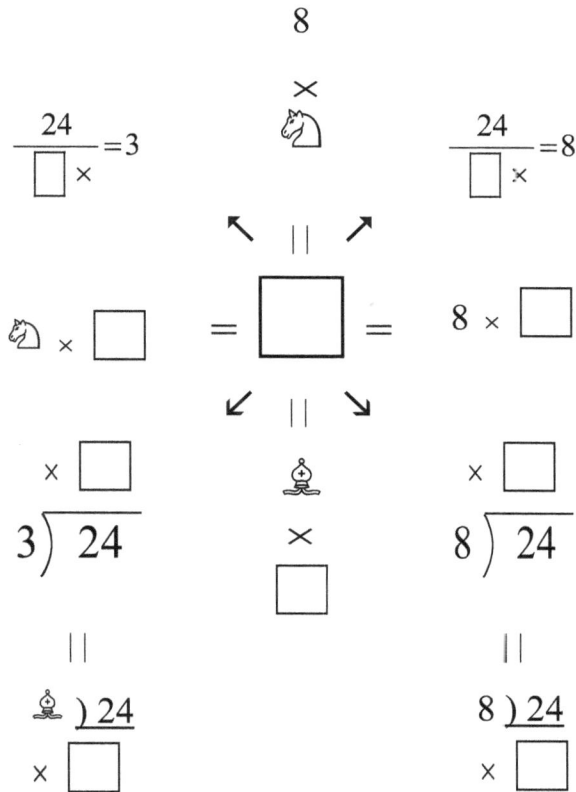

$$8$$
$$\times$$
♘

$$\frac{24}{\square \times} = 3 \qquad\qquad \frac{24}{\square \times} = 8$$

$$\nwarrow \ \| \ \nearrow$$

$$♘ \times \square \ = \ \boxed{} \ = \ 8 \times \square$$

$$\swarrow \ \| \ \searrow$$

$$\times \square \qquad\qquad ♗ \qquad\qquad \times \square$$

$$3\,\overline{)\,24} \qquad\qquad \times \qquad\qquad 8\,\overline{)\,24}$$

$$\square$$

$$\| \qquad\qquad\qquad\qquad \|$$

$$♗\,)\,24 \qquad\qquad\qquad 8\,)\,24$$

$$\times \square \qquad\qquad\qquad \times \square$$

2007 - 2017 © Frank Ho, Amanda Ho, All rights reserved.   www.homathchess.com

Different ways of writing multiplication (Learning division while doing multiplications)

2007 - 2017 © Frank Ho, Amanda Ho, All rights reserved.   www.homathchess.com

Different ways of writing multiplication (Learning division while doing multiplications)

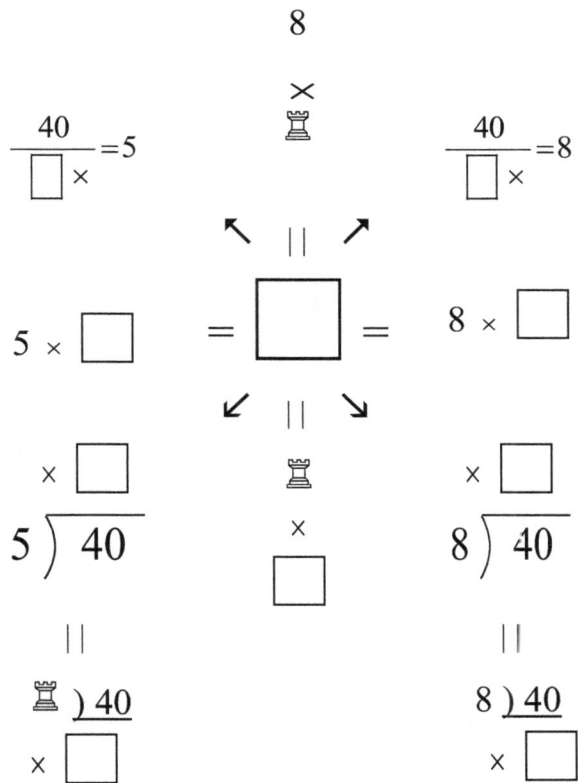

$$\frac{40}{\square \times} = 5 \qquad \frac{40}{\square \times} = 8$$

$$5 \times \square \quad = \quad \boxed{} \quad = \quad 8 \times \square$$

$$5 \overline{)\,40} \qquad\qquad 8 \overline{)\,40}$$

2007 - 2017 © Frank Ho, Amanda Ho, All rights reserved.　www.homathchess.com

Different ways of writing multiplication (Learning division while doing multiplications)

$$8$$
$$\times$$
$$6$$

$$\frac{48}{\Box\times} = 6 \qquad\qquad \frac{48}{\Box\times} = 8$$

$$6 \times \Box \quad = \quad \boxed{} \quad = \quad 8 \times \Box$$

$$\times \Box \qquad\qquad 6 \qquad\qquad \times \Box$$
$$6\,)\,\overline{48} \qquad\qquad \times \qquad\qquad 8\,)\,\overline{48}$$
$$\Box$$

$$6\,)\,\underline{48} \qquad\qquad\qquad 8\,)\,\underline{48}$$
$$\times \Box \qquad\qquad\qquad \times \Box$$

Ho Math Chess 何数棋谜 妈！我会棋谜式乘法啦！
Mom! I Learn Multiplication Using Math-Chess-Puzzles Connection!

Student's Name _____ Date _____

2007 - 2017 © Frank Ho, Amanda Ho, All rights reserved.   www.homathchess.com

Different ways of writing multiplication (Learning division while doing multiplications)

Ho Math Chess   何数棋谜 妈！我会棋谜式乘法啦！
Mom! I Learn Multiplication Using Math-Chess-Puzzles Connection!

Student's Name _____ Date _____

2007 - 2017 © Frank Ho, Amanda Ho, All rights reserved.   www.homathchess.com

Different ways of writing multiplication (Learning division while doing multiplications)

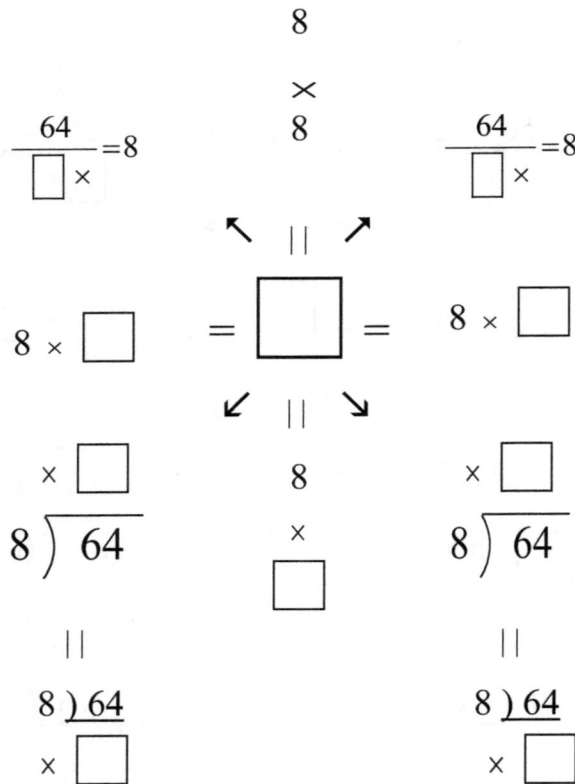

$$\frac{64}{\Box \times} = 8 \qquad\qquad \begin{array}{c} 8 \\ \times \\ 8 \end{array} \qquad\qquad \frac{64}{\Box \times} = 8$$

$$\nwarrow \; || \; \nearrow$$

$$8 \times \Box \qquad = \boxed{} = \qquad 8 \times \Box$$

$$\swarrow \; || \; \searrow$$

$$\begin{array}{c} \times \Box \\ 8\overline{)\,64} \end{array} \qquad \begin{array}{c} 8 \\ \times \\ \Box \end{array} \qquad \begin{array}{c} \times \Box \\ 8\overline{)\,64} \end{array}$$

$$||$$

$$\begin{array}{c} 8\,)\,64 \\ \times \Box \end{array} \qquad\qquad \begin{array}{c} 8\,)\,64 \\ \times \Box \end{array}$$

2007 - 2017 © Frank Ho, Amanda Ho, All rights reserved.   www.homathchess.com

Different ways of writing multiplication (Learning division while doing multiplications)

Ho Math Chess　何数棋谜　妈！我会棋谜式乘法啦！
Mom! I Learn Multiplication Using Math-Chess-Puzzles Connection!

Student's Name _____ Date _____

2007 - 2017 © Frank Ho, Amanda Ho, All rights reserved.　　www.homathchess.com

Counting 9s multiples

9, 18, 27, ☐,☐,☐,☐,☐,☐

Fill in the following ☐ with a number.

Sequence	1	2	♞	4	♜	6	7	8	♛
Add 9	☐	18	☐	36	☐	54	☐	72	☐

Sequence	♙	2	3	4	5	6	7	8	9
Add 9	9	☐	27	☐	45	☐	63	☐	81

Sequence	1	2	♞	4	♜	6	7	8	♛
Add 9	☐	18	☐	36	☐	54	☐	72	☐

Sequence	1	2	3	4	5	6	7	8	9
Add 9	9	☐	27	☐	45	☐	63	☐	81

Sequence	♙	2	♞	4	♜	6	7	8	♛
Add 9	☐	18	☐	36	☐	54	☐	72	☐

Sequence	1	2	3	4	♜	6	7	8	9
Add 9	9	☐	27	☐	45	☐	63	☐	81

Ho Math Chess   何数棋谜   妈！我会棋谜式乘法啦！
Mom! I Learn Multiplication Using Math-Chess-Puzzles Connection!

Student's Name _____ Date _____

2007 - 2017 © Frank Ho, Amanda Ho, All rights reserved.   www.homathchess.com

9 times

9 × 1 = ☐	Nine times one is ☐	1 × 9 = ☐	One times nine is ☐
9 × 2 = ☐	Nine times two is ☐	2 × 9 = ☐	Two times nine is ☐
9 × 3 = ☐	Nine times three is ☐	3 × 9 = ☐	Three times nine is ☐
9 × 4 = ☐	Nine times four is ☐	4 × 9 = ☐	Four times nine is ☐
9 × 5 = ☐	Nine times five is ☐	5 × 9 = ☐	Five times nine is ☐
9 × 6 = ☐	Nine times six is ☐	6 × 9 = ☐	Six times nine is ☐
9 × 7 = ☐	Nine times seven is ☐	7 × 9 = ☐	Seven times nine is ☐
9 × 8 = ☐	Nine times eight is ☐	8 × 9 = ☐	Eight times nine is ☐
9 × 9 = ☐	Nine times nine is ☐	9 × 9 = ☐	Nine times nine is ☐

```
    9          1          2          ♛          3
  X 1        X 9        X 9        X 2        X 9
   ☐          ☐         ☐☐        ☐☐        ☐☐


    5          9          9          8          9
  X ♛        X 6        X 7        X ♛        X 3
  ☐☐        ☐☐        ☐☐        ☐☐        ☐☐


    4          3          5          7          9
  X 9        X ♛        X 9        X 9        X ♛
  ☐☐        ☐☐        ☐☐        ☐☐        ☐☐
```

Ho Math Chess 何数棋谜 妈！我会棋谜式乘法啦！
Mom! I Learn Multiplication Using Math-Chess-Puzzles Connection!

Student's Name _____ Date _____

2007 - 2017 © Frank Ho, Amanda Ho, All rights reserved.   www.homathchess.com

9 × 1	1 × ♛	9 × 2	2 × 9	9 × 3
9 × 4	9 × 5	7 × 9	♛ × 8	9 × 6
♛ × 6	6 × 9	9 × 7	6 × ♛	9 × 8
9 × 5	6 × ♛	7 × 9	9 × 8	9 × 4
7 × 9	9 × 8	5 × 9	9 × ♝	6 × 9

2007 - 2017 © Frank Ho, Amanda Ho, All rights reserved.   www.homathchess.com

\

Oral practice

nine one nine	9 1 ☐	$\begin{array}{r} 1\ 1 \\ \times\quad 9 \\ \hline 99\ \square\square \end{array}$
nine two eighteen	♛ 2 ☐	$\begin{array}{r} 2\ 2 \\ \times\quad 9 \\ \hline 198\ \square\square\square \end{array}$
nine three twenty-seven	9 ♝ ☐	$\begin{array}{r} 3\ 3 \\ \times\quad 9 \\ \hline 297\ \square\square\square \end{array}$
nine four thirty-six	9 4 ☐	$\begin{array}{r} 4\ 4 \\ \times\quad ♛ \\ \hline 396\ \square\square\square \end{array}$
nine five forty-five	♛ ♜ ☐	$\begin{array}{r} {}^{4} \\ 5\ 5 \\ \times\quad 9 \\ \hline 495\ \square\square\square \end{array}$

2007 - 2017 © Frank Ho, Amanda Ho, All rights reserved.   www.homathchess.com

Oral practice

nine six fifty-four	9 6 ☐	$\overset{1}{6}6$ $\times$ ♛ 594 ☐☐☐
nine seven sixty-three	♛ 7 ☐	7 7 $\times$ 9 693 ☐☐☐
nine eight seventy-two	9 8 ☐	8 8 $\times$ 9 792 ☐☐☐
nine nine eighty-one	9 ♛ ☐	9 9 $\times$ 9 891 ☐☐☐
nine four thirty-six	9 4 ☐	4 4 $\times$ 9 396 ☐☐☐

2007 - 2017 © Frank Ho, Amanda Ho, All rights reserved.　　www.homathchess.com

Preparing for division

□	□	□	□	□
X 2	X ♗	X 4	X 5	X 6
18	27	36	45	54

□	□	□	□	□
X 7	X 8	X 9	X 3	X 4
63	72	81	27	36

□	□	□	□	□
X 9	X 9	X 5	X 2	X 7
45	18	45	18	63

□	□	□	□	□
X 4	X ♖	X 4	X ♗	X 8
36	45	36	27	72

□	□	□	□	□
X 4	X 5	X 6	X 2	X 7
36	45	54	18	63

Mom! I Learn Multiplication Using Math-Chess-Puzzles Connection!

Student's Name _____ Date _____

2007 - 2017 © Frank Ho, Amanda Ho, All rights reserved.   www.homathchess.com

Fill in ☐ with answer.

Times	Grouping	Addition
$9 \times \$1 = $ ☐	9 of ☐ $= 9$	$\$1 + \$1 + \$1 + \$1 + \$1 + \$1 + \$1 + \$1 + \$1 = $ ☐
$1 \times \$9 = $ ☐	1 of ☐ $= 9$	$\$9$

Fill in ☐ with answer.

Expression	Grouping	Addition
$9 \times \$2$	9 of ☐ $= 18$	$\$2 + \$2 + \$2 + \$2 + \$2 + \$2 + \$2 + \$2 + \$2 = $ ☐
$2 \times \$9$	2 of ☐ $= 18$	$\$9 + \$9 = $ ☐

$9 \times 1 = $ ☐ $= ♟ \times 9 = $ ☐	$1 \times 9 = $ ☐ $= 9 \times 1 = $ ☐
$9 \times $ ☐ $= 18 = 2 \times $ ☐ $= $ ☐	$2 \times $ ☐ $= 18 = 9 \times $ ☐ $= $ ☐

9 ♟ ☐	9 5 ☐	9 9 ☐	9 4 ☐	9 8 ☐
9 2 ☐	♛ 6 ☐	9 1 ☐	9 ♜ ☐	♛ 9 ☐
♛ 3 ☐	9 7 ☐	♛ 2 ☐	9 6 ☐	9 1 ☐
9 4 ☐	9 8 ☐	9 3 ☐	9 7 ☐	♛ 2 ☐

Mom! I Learn Multiplication Using Math-Chess-Puzzles Connection!

Student's Name _____ Date _____

2007 - 2017 © Frank Ho, Amanda Ho, All rights reserved.   www.homathchess.com

Fill in ☐ with answer.

Times	Grouping	Addition
$9 \times \$3 = $ ☐	9 of ☐ = 27	$\$3 + \$3 + \$3 + \$3 + \$3 + \$3 + \$3 + \$3 + \$3 = $ ☐
$3 \times \$9 = $ ☐	3 of ☐ = 27	$\$9 + \$9 + \$9 = $ ☐

Fill in ☐ with answer.

Times	Grouping	Addition
$9 \times \$4$	9 of ☐ = 36	$\$4 + \$4 + \$4 + \$4 + \$4 + \$4 + \$4 + \$4 + \$4 = $ ☐
$4 \times \$9$	4 of ☐ = 36	$\$9 + \$9 + \$9 + \$9 = $ ☐

$9 \times 3 = $ ☐ $ = 3 \times 9 = $ ☐	$3 \times 9 = $ ☐ $ = 9 \times 3 = $ ☐
$9 \times $ ☐ $ = 36 = 4 \times $ ☐ $ = $ ☐	$4 \times $ ☐ $ = 36 = 9 \times $ ☐ $ = $ ☐

♛ 1 ☐	9 ♜ ☐	♛ 9 ☐	9 4 ☐	9 8 ☐
9 2 ☐	9 6 ☐	9 1 ☐	9 5 ☐	9 ♛ ☐
9 ♗ ☐	♛ 7 ☐	9 2 ☐	♛ 6 ☐	9 1 ☐
9 4 ☐	9 8 ☐	9 3 ☐	9 7 ☐	9 2 ☐

Fill in [ ] with answer.

Times	Grouping	Addition
$9 \times \$5 = \square$	9 of $\square = 45$	$\$5 + \$5 + \$5 + \$5 + \$5 + \$5 + \$5 + \$5 + \$5 = \square$
$5 \times \$9 = \square$	5 of $\square = 45$	$\$9 + \$9 + \$9 + \$9 + \$9 = \square$

Times	Grouping	Addition
$9 \times \$6 = \square$	9 of $\square = 54$	$\$6 + \$6 + \$6 + \$6 + \$6 + \$6 + \$6 + \$6 + \$6 = \square$
$6 \times \$9 = \square$	6 of $\square = 54$	$\$9 + \$9 + \$9 + \$9 + \$9 + \$9 = \square$

$9 \times 5 = \square = ♖ \times 9 = \square$	$5 \times 9 = \square = 9 \times 5 = \square$
$9 \times \square = 54 = 6 \times \square = \square$	$6 \times \square = 54 = 9 \times \square = \square$

♛ 1 $\square$	9 5 $\square$	♛ 9 $\square$	9 4 $\square$	9 8 $\square$
9 2 $\square$	9 6 $\square$	9 ♙ $\square$	9 ♖ $\square$	9 9 $\square$
9 3 $\square$	♛ 7 $\square$	9 2 $\square$	♛ 6 $\square$	9 ♙ $\square$
9 4 $\square$	9 8 $\square$	9 3 $\square$	9 7 $\square$	♛ 2 $\square$

2007 - 2017 © Frank Ho, Amanda Ho, All rights reserved.   www.homathchess.com

Fill in _____ and ☐ with answers.

Times	Grouping	Addition
$9 \times \$7 =$ ☐	9 of ☐ = 63	$\$7 + \$7 + \$7 + \$7 + \$7 + \$7 + \$7 + \$7 + \$7 =$ ☐
$7 \times \$9 =$ ☐	7 of ☐ = 63	$\$9 + \$9 + \$9 + \$9 + \$9 + \$9 + \$9 =$ ☐

Times	Grouping	Addition
$9 \times \$8 =$ ☐	9 of ☐ = 72	$\$8 + \$8 + \$8 + \$8 + \$8 + \$8 + \$8 + \$8 + \$8 =$ ☐
$8 \times \$9 =$ ☐	8 of ☐ = 72	$\$9 + \$9 + \$9 + \$9 + \$9 + \$9 + \$9 + \$9 =$ ☐

$9 \times 7 =$ ☐ $= 7 \times 9 =$ ☐	$7 \times 9 =$ ☐ $= 9 \times 7 =$ ☐
$9 \times$ ☐ $= 72 = 8 \times$ ☐ $=$ ☐	$8 \times$ ☐ $= 72 = 9 \times$ ☐ $=$ ☐

9 1 ☐	9 5 ☐	9 ♛ ☐	9 4 ☐	♛ 8 ☐
♛ 2 ☐	9 6 ☐	9 1 ☐	9 ♜ ☐	9 9 ☐
9 3 ☐	9 7 ☐	9 2 ☐	♛ 6 ☐	9 1 ☐
9 4 ☐	♛ 8 ☐	9 ♗ ☐	9 7 ☐	♛ 2 ☐

Ho Math Chess 何数棋谜 妈!我会棋谜式乘法啦!
Mom! I Learn Multiplication Using Math-Chess-Puzzles Connection!

Student's Name _____ Date _____

2007 - 2017 © Frank Ho, Amanda Ho, All rights reserved.   www.homathchess.com

Fill in ☐ with answer.

Times	Grouping	Addition
$9 \times \$9 = $ ☐	9 of ☐ $= 72$	$\$9 + \$9 + \$9 + \$9 + \$9 + \$9 + \$9 + \$9 + \$9 = $ ☐
$9 \times \$9 = $ ☐	8 of ☐ $= 72$	$\$9 + \$9 + \$9 + \$9 + \$9 + \$9 + \$9 + \$9 + \$9 = $ ☐

Times	Grouping	Addition
$9 \times \$8 = $ ☐	9 of ☐ $= 72$	$\$8 + \$8 + \$8 + \$8 + \$8 + \$8 + \$8 + \$8 + \$8 = $ ☐
$8 \times \$9 = $ ☐	8 of ☐ $= 72$	$\$9 + \$9 + \$9 + \$9 + \$9 + \$9 + \$9 + \$9 = $ ☐

$9 \times 9 = $ ☐ $= 9 \times $ ♛ $= $ ☐	$9 \times 9 = $ ☐ $= $ ♛ $\times 9 = $ ☐
$9 \times $ ☐ $= 81 = 9 \times $ ☐ $= $ ☐	$9 \times $ ☐ $= 81 = 9 \times $ ☐ $= $ ☐

$9\ 1$ ☐	$9\ 5$ ☐	♛ 9 ☐	$9\ 4$ ☐	$9\ 8$ ☐
$9\ 2$ ☐	♛ 6 ☐	$9\ 1$ ☐	9 ♖ ☐	9 ♛ ☐
$9\ 3$ ☐	$9\ 7$ ☐	$9\ 2$ ☐	♛ 6 ☐	$9\ 1$ ☐
♛ 4 ☐	$9\ 8$ ☐	9 ♗ ☐	$9\ 7$ ☐	$9\ 2$ ☐

2007 - 2017 © Frank Ho, Amanda Ho, All rights reserved.　　www.homathchess.com

Preparing for division

☐ X 9 = 9	X ☐ 9)9	☐)9___ X 9
☐ X ♛ = 18	X ☐ 9)18	☐)18___ X ♛
☐ X 9 = 27	X ☐ 9)27	☐)27___ X 9
☐ X ♛ = 36	X ☐ 9)36	☐)36___ X ♛
☐ X 9 = 45	X ☐ 9)45	☐)45___ X 9
☐ X ♛ = 54	X ☐ 9)54	☐)54___ X ♛
☐ X 9 = 63	X ☐ 9)63	☐)63___ X 9

Preparing for division

☐ X 9 = 72	X ☐ 9)72	☐)72 X ♛
☐ X 9 = 81	X ☐ 9)81	☐)81 X 9
☐ X ♛ = 9	X ☐ 9)9	☐)9 X 9
☐ X 9 = 18	X ☐ 9)18	☐)18 X ♛
☐ X 9 = 27	X ☐ 9)27	☐)27 X 9
☐ X ♛ = 36	X ☐ 9)36	☐)36 X 9
☐ X 9 = 45	X ☐ 9)45	☐)45 X ♛

2007 - 2017 © Frank Ho, Amanda Ho, All rights reserved.　www.homathchess.com

Cross multiplication

12　　　　12 ↖ ↗ $\frac{6}{2} = \frac{6}{2}$	☐　　　18 ↖ ↗ $\frac{9}{9} = \frac{2}{2}$	☐　　　27 ↖ ↗ $\frac{9}{9} = \frac{3}{3}$	☐　　　36 ↖ ↗ $\frac{9}{9} = \frac{4}{4}$
☐　　　45 ↖ ↗ $\frac{9}{9} = \frac{5}{5}$	☐　　　54 ↖ ↗ $\frac{9}{9} = \frac{6}{6}$	☐　　　63 ↖ ↗ $\frac{9}{9} = \frac{7}{7}$	☐　　　81 ↖ ↗ $\frac{9}{9} = \frac{9}{9}$
☐　　　45 ↖ ↗ $\frac{9}{9} = \frac{5}{5}$	☐　　　36 ↖ ↗ $\frac{9}{9} = \frac{4}{4}$	☐　　　63 ↖ ↗ $\frac{9}{9} = \frac{7}{7}$	☐　　　72 ↖ ↗ $\frac{9}{9} = \frac{8}{8}$
☐　　　54 ↖ ↗ $\frac{9}{9} = \frac{6}{6}$	☐　　　72 ↖ ↗ $\frac{9}{9} = \frac{8}{8}$	☐　　　81 ↖ ↗ $\frac{9}{9} = \frac{9}{9}$	☐　　　27 ↖ ↗ $\frac{9}{9} = \frac{3}{3}$

Ho Math Chess　　何数棋谜　　妈！我会棋谜式乘法啦！
Mom! I Learn Multiplication Using Math-Chess-Puzzles Connection!

Student's Name _____ Date _____

2007 - 2017 © Frank Ho, Amanda Ho, All rights reserved.　　www.homathchess.com

Different ways of writing multiplication (Learning division while doing multiplications)

Ho Math Chess 何数棋谜 妈！我会棋谜式乘法啦！
Mom! I Learn Multiplication Using Math-Chess-Puzzles Connection!

Student's Name _____ Date _____

2007 - 2017 © Frank Ho, Amanda Ho, All rights reserved.   www.homathchess.com

Different ways of writing multiplication (Learning division while doing multiplications)

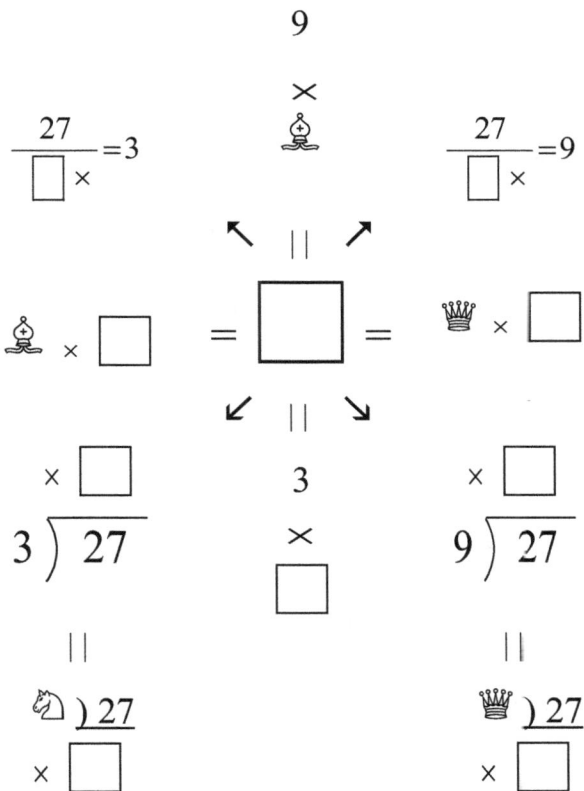

$$9$$
$$\times$$

$$\frac{27}{\boxed{}\times} = 3 \qquad \qquad ♗ \qquad \qquad \frac{27}{\boxed{}\times} = 9$$

$$\nwarrow \ \| \ \nearrow$$

$$♗ \times \boxed{} = \boxed{} = ♕ \times \boxed{}$$

$$\swarrow \ \| \ \searrow$$

$$\times \boxed{} \qquad \qquad 3 \qquad \qquad \times \boxed{}$$

$$3 \overline{)\,27} \qquad \qquad \times \qquad \qquad 9 \overline{)\,27}$$

$$\boxed{}$$

$$\| \qquad \qquad \qquad \qquad \|$$

$$♘ \,)\,27 \qquad \qquad ♕ \,)\,27$$

$$\times \boxed{} \qquad \qquad \times \boxed{}$$

Ho Math Chess　何数棋谜　妈!我会棋谜式乘法啦!
Mom! I Learn Multiplication Using Math-Chess-Puzzles Connection!

Student's Name _____ Date _____

2007 - 2017 © Frank Ho, Amanda Ho, All rights reserved.　www.homathchess.com

Different ways of writing multiplication (Learning division while doing multiplications)

$$9 \times 4$$

$$\frac{36}{\square \times} = 4 \qquad \frac{36}{\square \times} = 9$$

$$4 \times \square = \boxed{} = \square \times ♛$$

$$4 \overline{)\,36} \qquad 4 \times \square \qquad 9 \overline{)\,36}$$

$$4 \,)\,36 \times \square \qquad ♛ \,)\,36 \times \square$$

Ho Math Chess   何数棋谜   妈！我会棋谜式乘法啦！
Mom! I Learn Multiplication Using Math-Chess-Puzzles Connection!

Student's Name _____ Date _____

2007 - 2017 © Frank Ho, Amanda Ho, All rights reserved.   www.homathchess.com

Different ways of writing multiplication (Learning division while doing multiplications)

Different ways of writing multiplication (Learning division while doing multiplications)

$$9 \times 6$$

$$\frac{54}{\square \times} = 6$$

$$\frac{54}{\square \times} = 9$$

$$6 \times \square = \boxed{} = ♛ \times \square$$

$$\times \square$$
$$6 \overline{)\, 54}$$

$$6 \times \square$$

$$\times \square$$
$$9 \overline{)\, 54}$$

$$6 \underline{)\, 54}$$
$$\times \square$$

$$♛ \underline{)\, 54}$$
$$\times \square$$

2007 - 2017 © Frank Ho, Amanda Ho, All rights reserved.   www.homathchess.com

Different ways of writing multiplication (Learning division while doing multiplications)

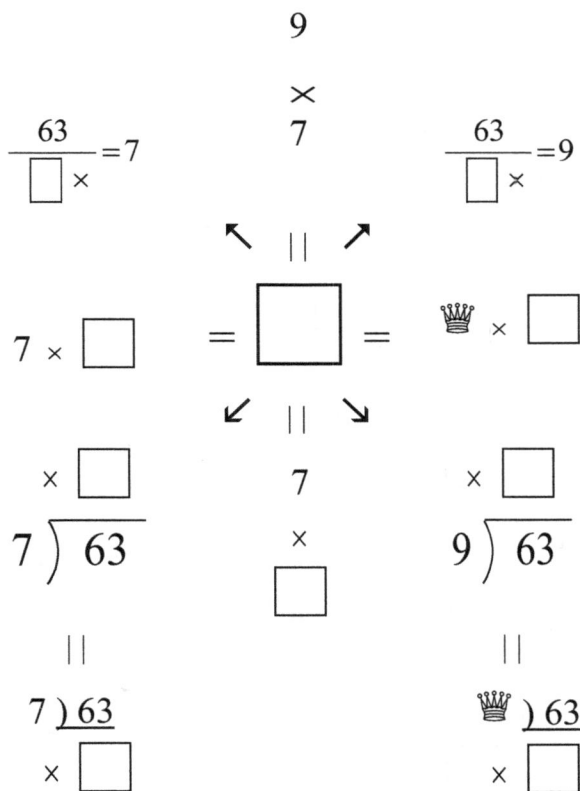

$$9$$
$$\times$$
$$7$$

$$\frac{63}{\boxed{}\times}=7$$

$$\frac{63}{\boxed{}\times}=9$$

$$\nwarrow \ \| \ \nearrow$$

$$7 \times \boxed{} \quad = \quad \boxed{} \quad = \quad ♛ \times \boxed{}$$

$$\swarrow \ \| \ \searrow$$

$$\times \boxed{}$$
$$7\overline{)\,63}$$

$$7$$
$$\times$$
$$\boxed{}$$

$$\times \boxed{}$$
$$9\overline{)\,63}$$

$$\|$$

$$\|$$

$$7\,\overline{)\,63}$$
$$\times \boxed{}$$

$$♛\,\overline{)\,63}$$
$$\times \boxed{}$$

2007 - 2017 © Frank Ho, Amanda Ho, All rights reserved.   www.homathchess.com

Different ways of writing multiplication (Learning division while doing multiplications)

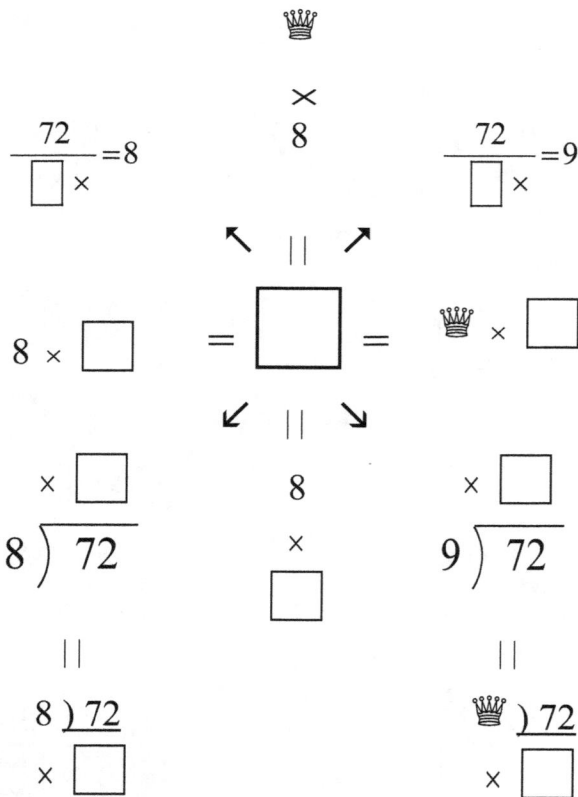

2007 - 2017 © Frank Ho, Amanda Ho, All rights reserved.   www.homathchess.com

Different ways of writing multiplication (Learning division while doing multiplications)

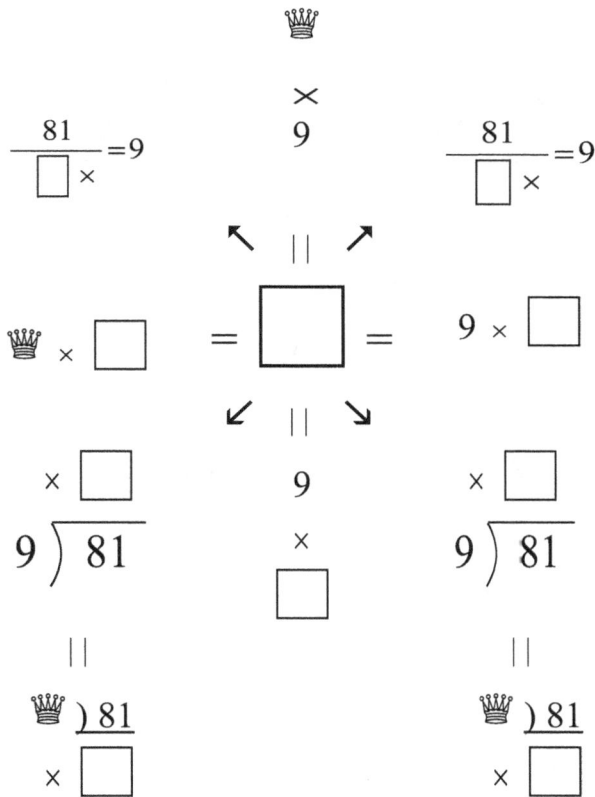

$$\frac{81}{\square \times} = 9 \qquad \begin{array}{c} \text{♛} \\ \times \\ 9 \end{array} \qquad \frac{81}{\square \times} = 9$$

$$\text{♛} \times \square \;=\; \boxed{} \;=\; 9 \times \square$$

$$\begin{array}{c} \times \;\square \\ 9\,\overline{)\,81} \end{array} \qquad \begin{array}{c} 9 \\ \times \\ \square \end{array} \qquad \begin{array}{c} \times \;\square \\ 9\,\overline{)\,81} \end{array}$$

$$\text{♛}\,)\,81 \qquad\qquad \text{♛}\,)\,81$$
$$\times\;\square \qquad\qquad\qquad \times\;\square$$

195

Product using image process

White has ___9____ product points.

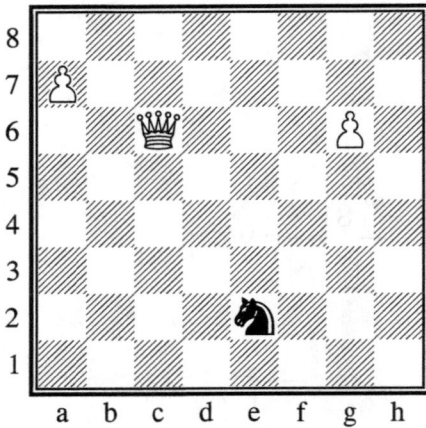

White has ___27____ product points.

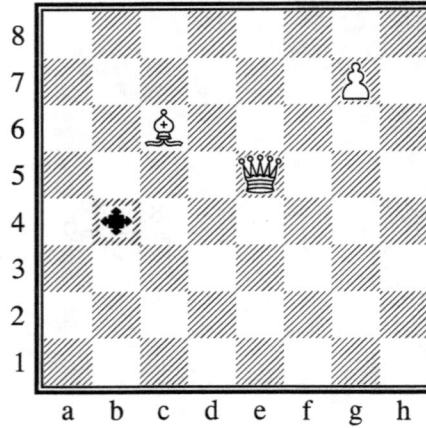

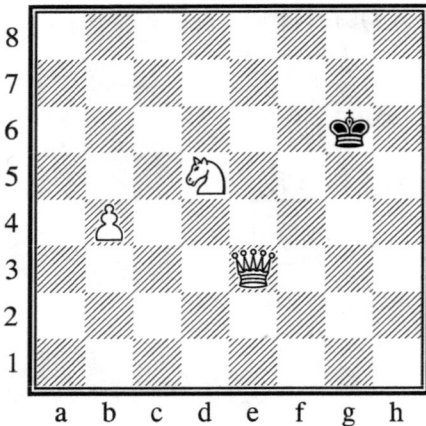

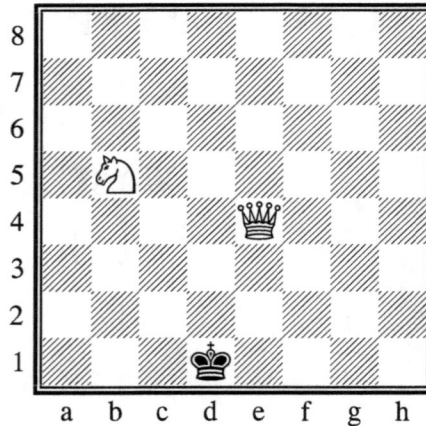

White has ____27____ product points.

White has ___27____ product points.

Ho Math Chess 何数棋谜 妈！我会棋谜式乘法啦！
Mom! I Learn Multiplication Using Math-Chess-Puzzles Connection!

Student's Name _____ Date _____

2007 - 2017 © Frank Ho, Amanda Ho, All rights reserved.   www.homathchess.com

Product using image process

White has ___27___ product points.

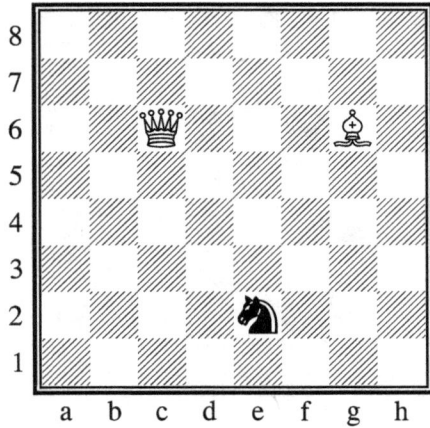

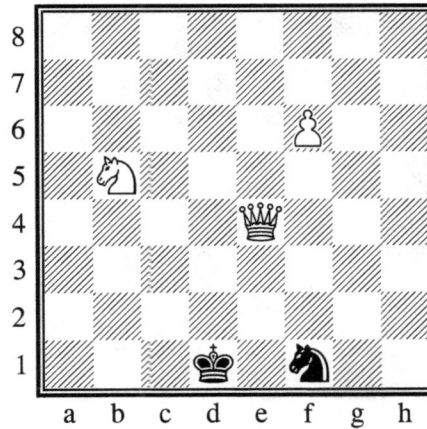

White has ___81___ product points.

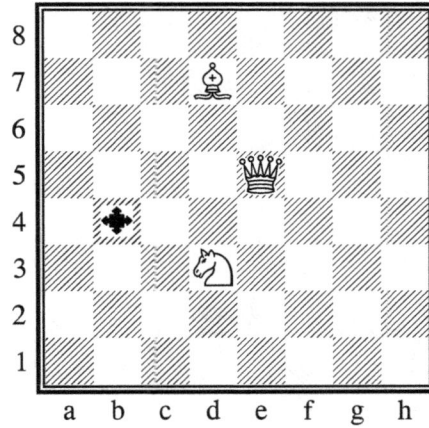

White has __135____ product points.

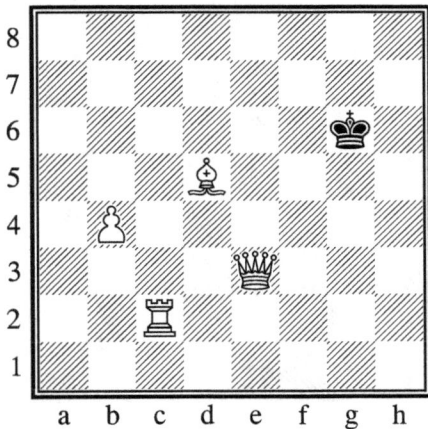

White has __27____ product points.

2007 - 2017 © Frank Ho, Amanda Ho, All rights reserved.   www.homathchess.com

Product using image process

White has __27____ product points.

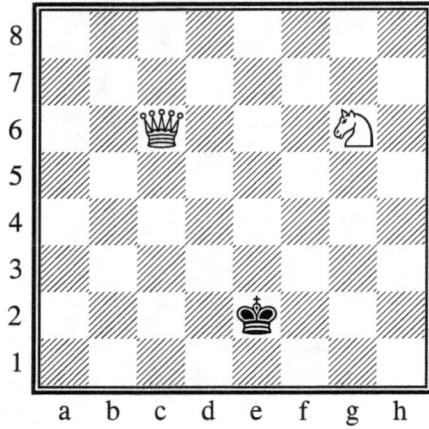

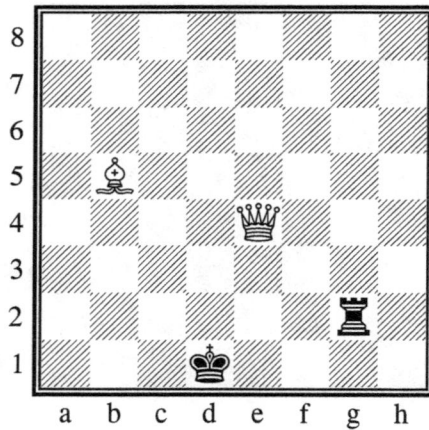

White has ___45___ product points.

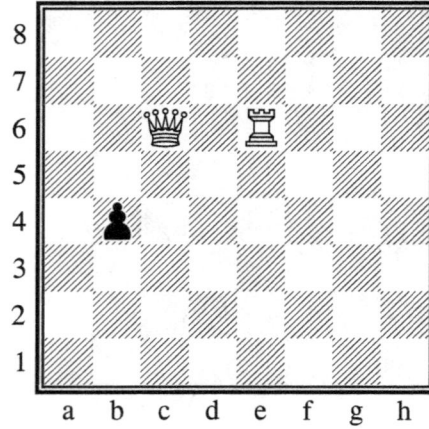

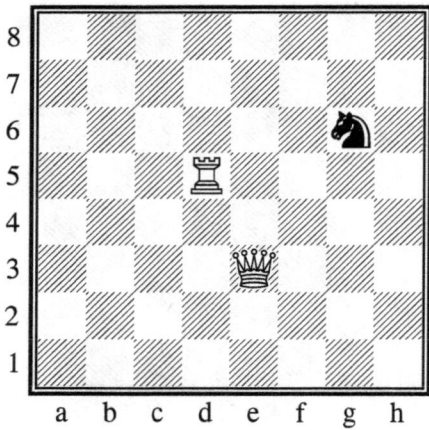

White has ___45___ product points.

White has __27____ product points.

Mom! I Learn Multiplication Using Math-Chess-Puzzles Connection!

Student's Name _____ Date _____

2007 - 2017 © Frank Ho, Amanda Ho, All rights reserved.　www.homathchess.com

Product using image process

White has ___27___ product points.

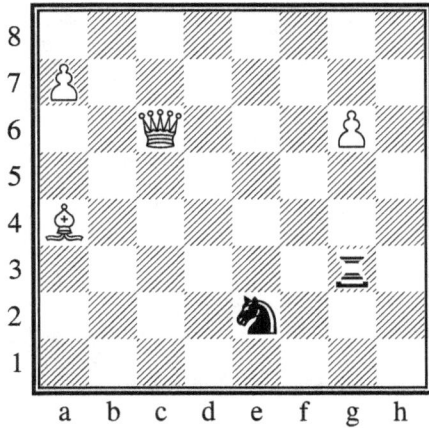

White has ___135___ product points.

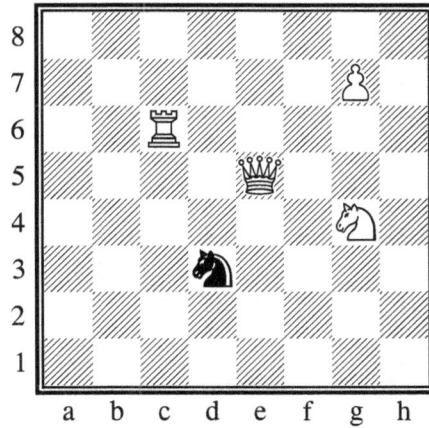

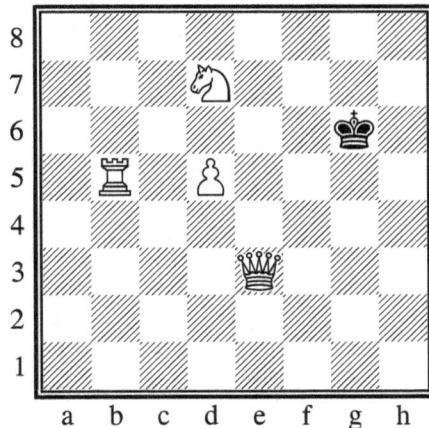

White has ___135___ product points.

White has __81____ product points.

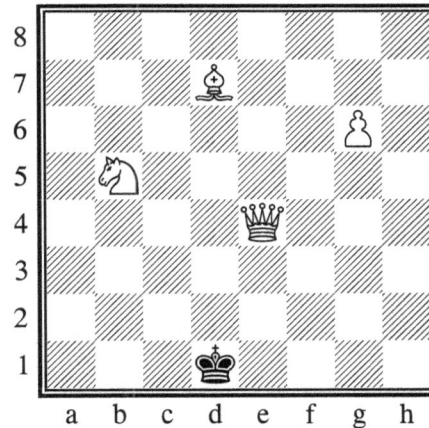

2007 - 2017 © Frank Ho, Amanda Ho, All rights reserved.   www.homathchess.com

Product using image process

White has ___81___ product points.

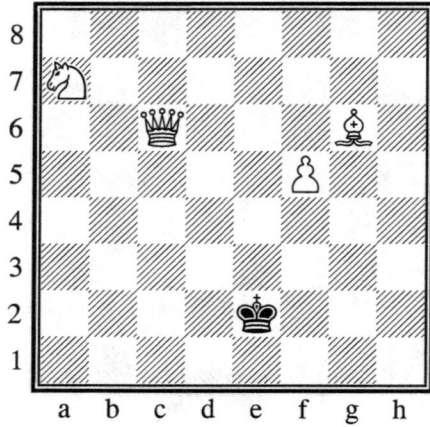

White has ___243___ product points.

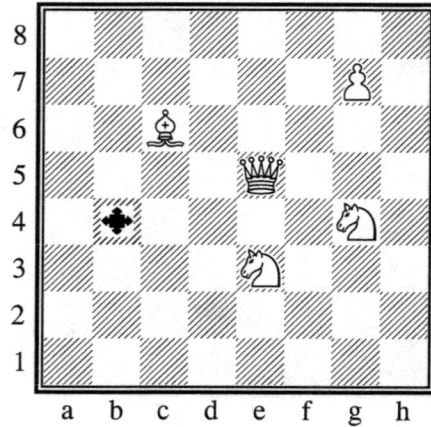

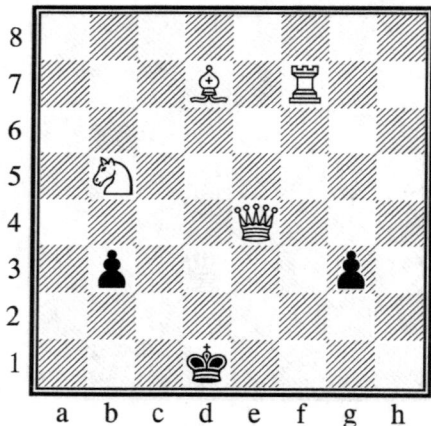

White has ___135___ product points.

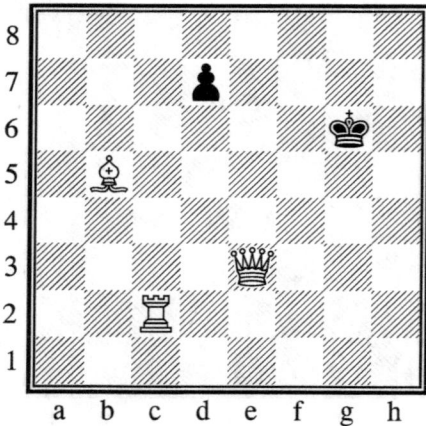

White has ___243___ product points.

2007 - 2017 © Frank Ho, Amanda Ho, All rights reserved.   www.homathchess.com

Product using image process

White has ___81___ product points.

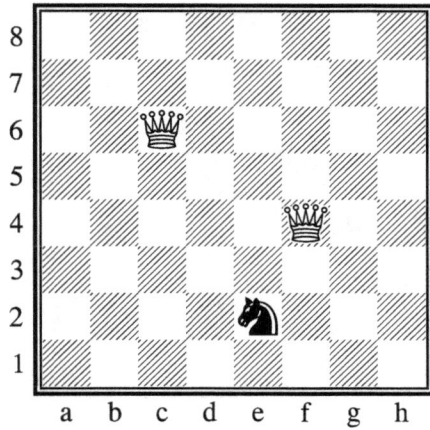

White has ___243___ product points.

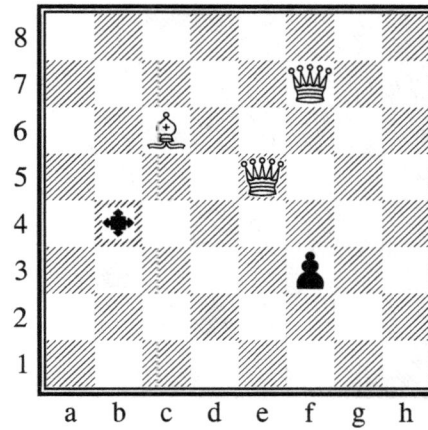

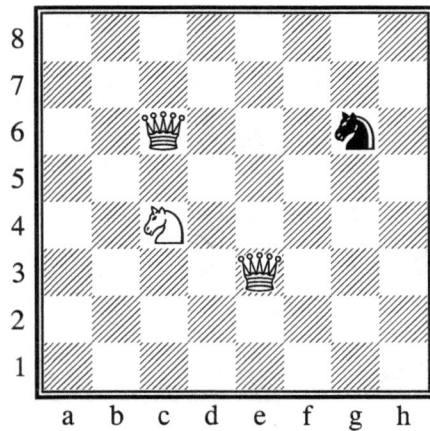

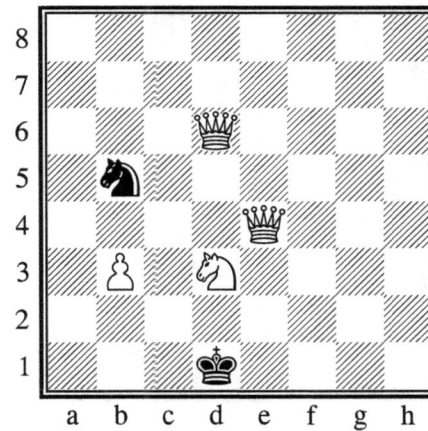

White has ___243___ product points.

White has ___243___ product points.

Ho Math Chess 何数棋谜 妈！我会棋谜式乘法啦！
Mom! I Learn Multiplication Using Math-Chess-Puzzles Connection!

Student's Name _____ Date _____

2007 - 2017 © Frank Ho, Amanda Ho, All rights reserved.   www.homathchess.com

Product using image process

White has ___243___ product points.

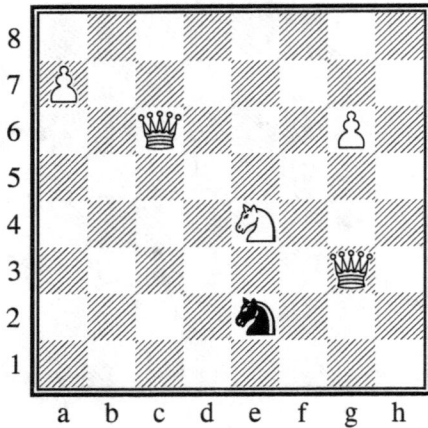

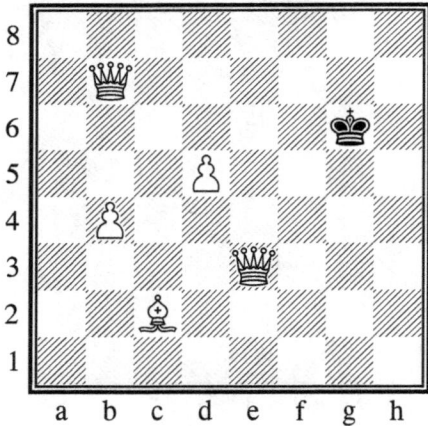

White has ___729___ product points.

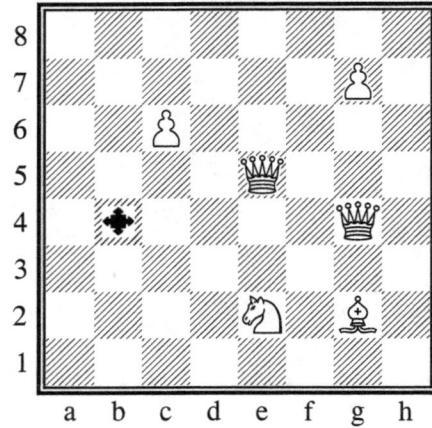

White has ___243___ product points.

White has __1215____ product points.

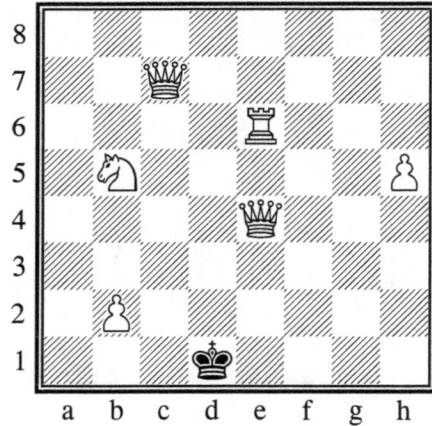

2007 - 2017 © Frank Ho, Amanda Ho, All rights reserved.   www.homathchess.com

Product using image process

White has ___405___ product points.

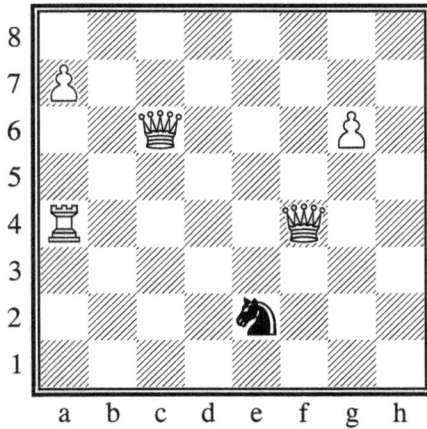

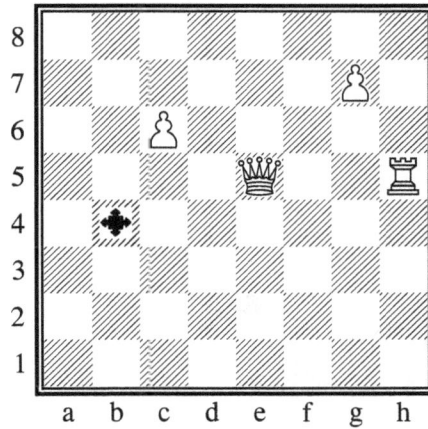

White has ___45___ product points.

White has ___27___ product points.

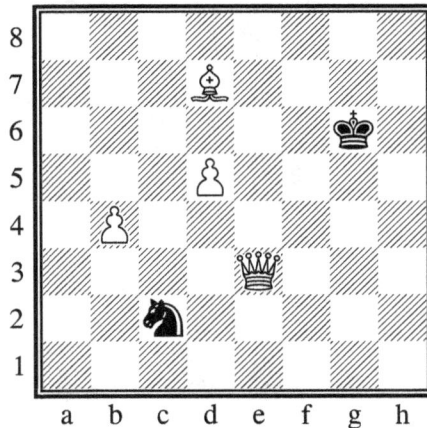

White has ___135___ product points.

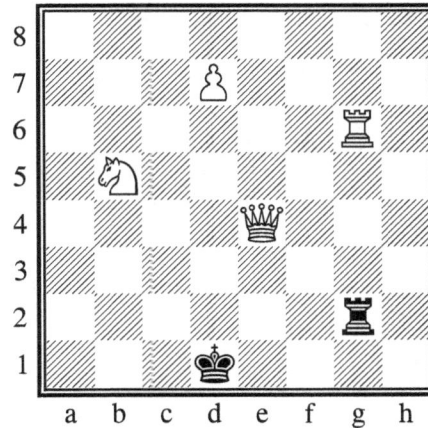

Ho Math Chess   何数棋谜   妈！我会棋谜式乘法啦！
Mom! I Learn Multiplication Using Math-Chess-Puzzles Connection!

Student's Name _____ Date _____

2007 - 2017 © Frank Ho, Amanda Ho, All rights reserved.   www.homathchess.com

Product using image process

White has ___45___ product points.

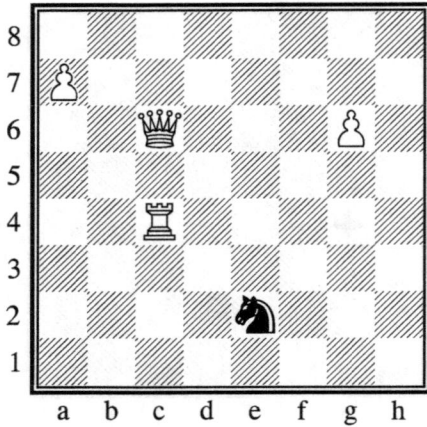

White has ___45___ product points.

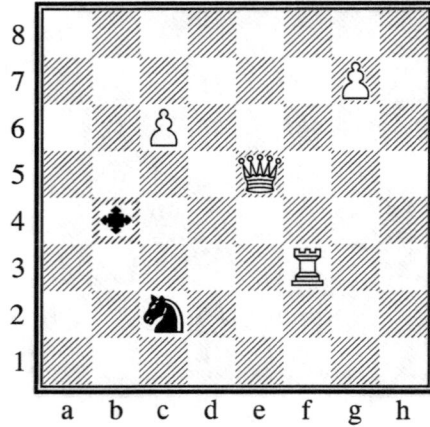

White has ___45___ product points.

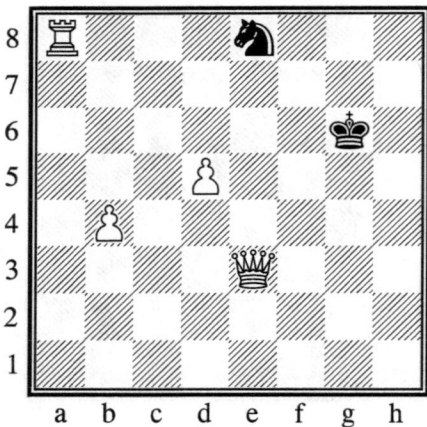

White has ___135___ product points.

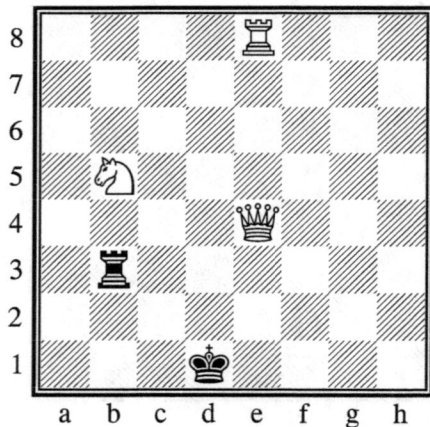

Ho Math Chess 何数棋谜 妈！我会棋谜式乘法啦！
Mom! I Learn Multiplication Using Math-Chess-Puzzles Connection!

Student's Name _____ Date _____

2007 - 2017 © Frank Ho, Amanda Ho, All rights reserved.   www.homathchess.com

Multiplying by relating

↓× ♛ ×↓ ♟ × 2 9☐ ♟ 18☐ × ♟ 9☐	↓× ♛ ×↓ 2 × ♝ 18☐ 2 27☐ × ♟ 18☐	↓× ♛ ×↓ ♝ × 4 27☐ ♝ 36☐ × ♟ 27☐
↓× 9 ×↓ 4 × ♜ 36☐ 4 45☐ × 1 36☐	↓× ♛ ×↓ ♜ × 6 45☐ ♜ 54☐ × 1 45☐	↓× 9 ×↓ 6 × 7 54☐ 6 63☐ × ♟ 54☐
↓× ♛ ×↓ 7 × 8 63☐ 7 72☐ × ♟ 63☐	↓× 9 ×↓ 8 × ♛ 72☐ 8 81☐ × 1 72☐	↓× 8 ×↓ 8 × ♛ 64☐ 8 72☐ × ♟ 64☐

205

2007 - 2017 © Frank Ho, Amanda Ho, All rights reserved.　　www.homathchess.com

Multiplying by relating

Ho Math Chess   何数棋谜   妈！我会棋谜式乘法啦！
Mom! I Learn Multiplication Using Math-Chess-Puzzles Connection!

Student's Name _____   Date _____

2007 - 2017 © Frank Ho, Amanda Ho, All rights reserved.   www.homathchess.com

Multiplying by relating

2007 - 2017 © Frank Ho, Amanda Ho, All rights reserved.   www.homathchess.com

Multiplying by relating

↓× ♞ ×↓ ♙ × 2 3☐ ♙ 6☐ × ♙ 3☐	↓× ♞ ×↓ 2 × ♝ 6☐ 2 9☐ × ♙ 6☐	↓× ♞ ×↓ ♝ × 4 9☐ ♝ 12☐ × ♙ 9☐
↓× ♞ ×↓ 4 × ♜ 12☐ 4 15☐ × 1 12☐	↓× ♞ ×↓ ♜ × 6 15☐ ♜ 18☐ × 1 15☐	↓× ♞ ×↓ 6 × 7 18☐ 6 21☐ × ♙ 18☐
↓× ♞ ×↓ 7 × 8 21☐ 7 24☐ × ♙ 21☐	↓× ♞ ×↓ 8 × ♛ 24☐ 8 27☐ × 1 24☐	↓× ♞ ×↓ 8 × ♛ 24☐ 8 27☐ × ♙ 24☐

Ho Math Chess   何数棋谜   妈！我会棋谜式乘法啦！
Mom! I Learn Multiplication Using Math-Chess-Puzzles Connection!

Student's Name _____ Date _____

2007 - 2017 © Frank Ho, Amanda Ho, All rights reserved.   www.homathchess.com

Multiplying by relating

2007 - 2017 © Frank Ho, Amanda Ho, All rights reserved.   www.homathchess.com

Multiplying by relating

2007 - 2017 © Frank Ho, Amanda Ho, All rights reserved.   www.homathchess.com

Multiplying by relating

Ho Math Chess 何数棋谜 妈！我会棋谜式乘法啦！
Mom! I Learn Multiplication Using Math-Chess-Puzzles Connection!

Student's Name _____ Date _____

2007 - 2017 © Frank Ho, Amanda Ho, All rights reserved.   www.homathchess.com

Multiplying by relating

Ho Math Chess 何数棋谜 妈！我会棋谜式乘法啦！
Mom! I Learn Multiplication Using Math-Chess-Puzzles Connection!

Student's Name _____ Date _____

2007 - 2017 © Frank Ho, Amanda Ho, All rights reserved.   www.homathchess.com

Multiplying by relating

⌐× ♗ ×⌐ ♗ × 2 9☐ ♗ 6☐ × ♙ 9☐	⌐× ♗ ×⌐ ♘ × ♝ 9☐ 2 9☐ × ♙ 6☐	⌐× ♗ ×⌐ ♝ × 4 9☐ ♝ 12☐ × ♙ 9☐
⌐× 9 ×⌐ 4 × ♖ 36☐ ♗ 45☐ × 1 27☐	⌐× ♕ ×⌐ ♖ × 6 45☐ ♗ 54☐ × 1 27☐	⌐× 9 ×⌐ 6 × 7 54☐ ♗ 63☐ × ♙ 27☐
⌐× ♕ ×⌐ 7 × 8 63☐ ♗ 72☐ × ♙ 27☐	⌐× 9 ×⌐ 8 × ♕ 72☐ ♗ 36☐ × 1 27☐	⌐× 8 ×⌐ 8 × ♕ 64☐ ♗ 72☐ × ♙ 24☐

Ho Math Chess　何数棋谜　妈！我会棋谜式乘法啦！
Mom! I Learn Multiplication Using Math-Chess-Puzzles Connection!

Student's Name _____ Date _____

2007 - 2017 © Frank Ho, Amanda Ho, All rights reserved.　　www.homathchess.com

What is the product of all attacking pieces on the ▨ or ☒?

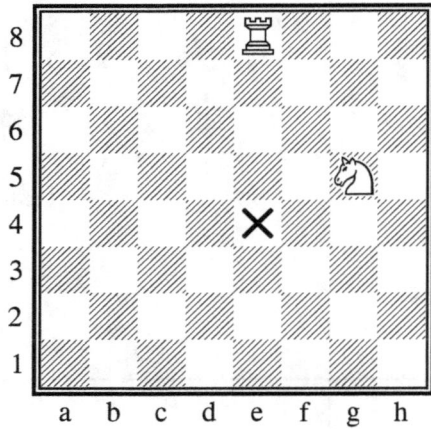

Answer _____15_____

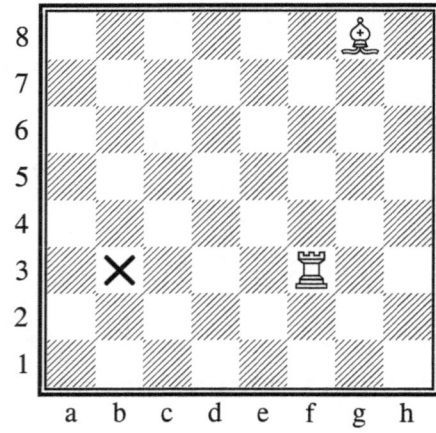

Answer _____15_____

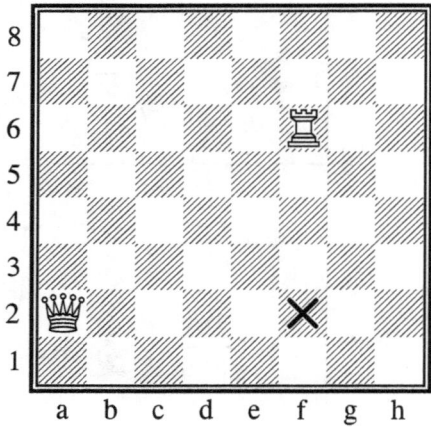

Answer _____45_____

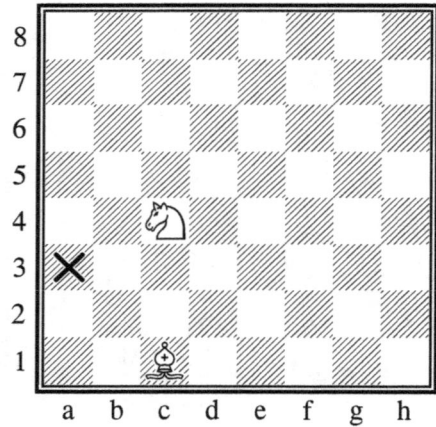

Answer _____9_____

Ho Math Chess　何数棋谜　妈！我会棋谜式乘法啦！
Mom! I Learn Multiplication Using Math-Chess-Puzzles Connection!

Student's Name _____ Date _____

2007 - 2017 © Frank Ho, Amanda Ho, All rights reserved.　www.homathchess.com

What is the product of all attacking pieces on the ✖ or ☒?

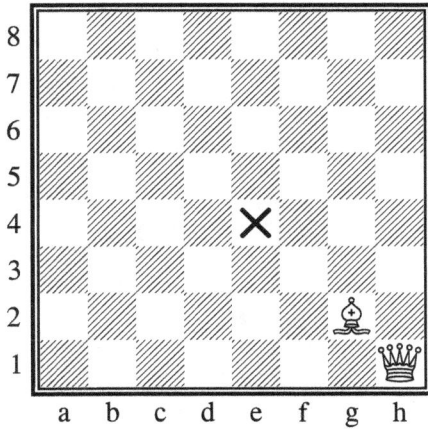

Answer _____27_____

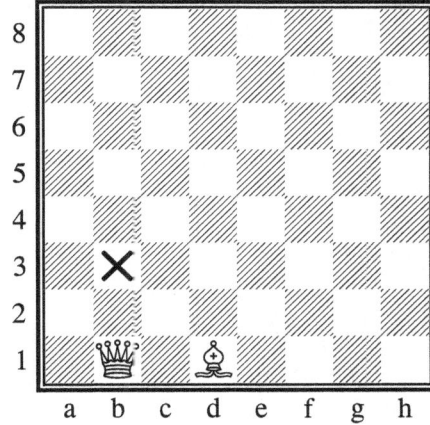

Answer _____27_____

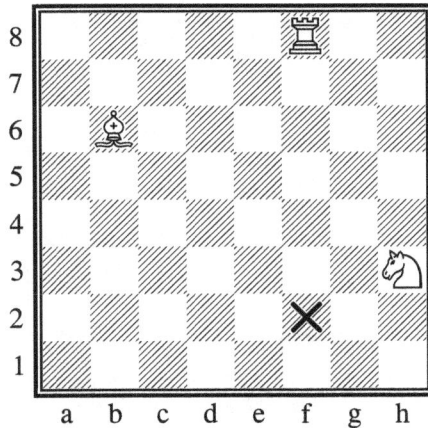

Answer _____45_____

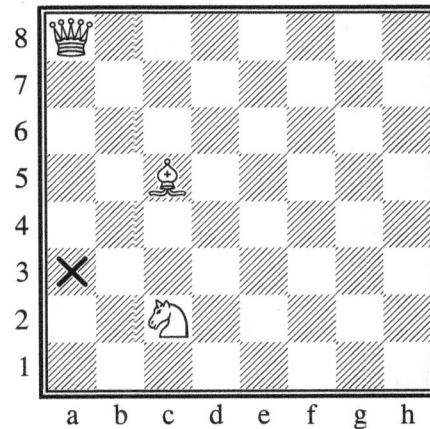

Answer _____81_____

Ho Math Chess 何数棋谜 妈！我会棋谜式乘法啦！
Mom! I Learn Multiplication Using Math-Chess-Puzzles Connection!

Student's Name _____ Date _____

2007 - 2017 © Frank Ho, Amanda Ho, All rights reserved.   www.homathchess.com

What is the product of all attacking pieces on the ✖ or ☒?

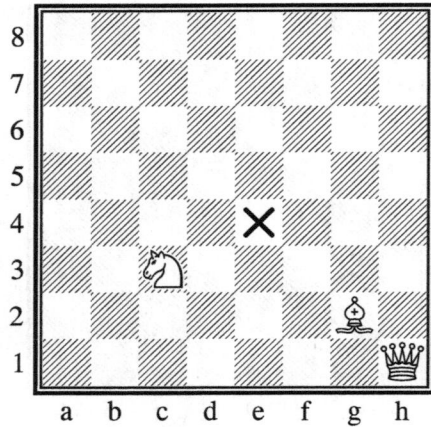

Answer _____81_____

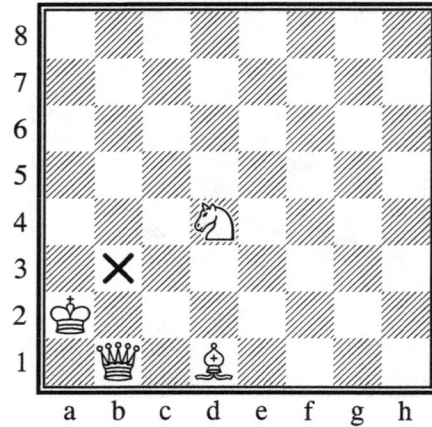

Answer _____0_____

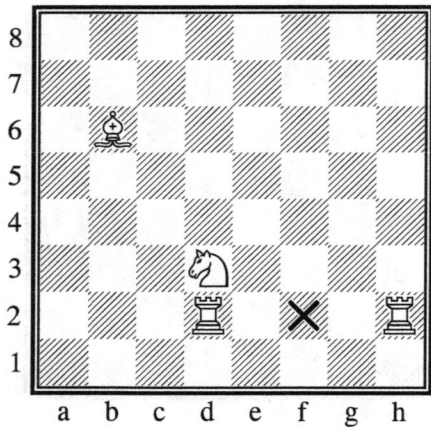

Answer _____225_____

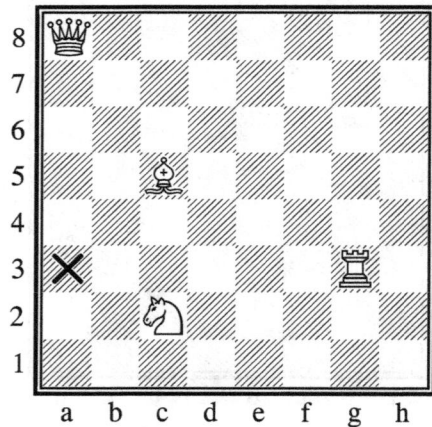

Answer _____405_____

2007 - 2017 © Frank Ho, Amanda Ho, All rights reserved.   www.homathchess.com

What is the product of all attacking pieces on the ✖ or ☒?

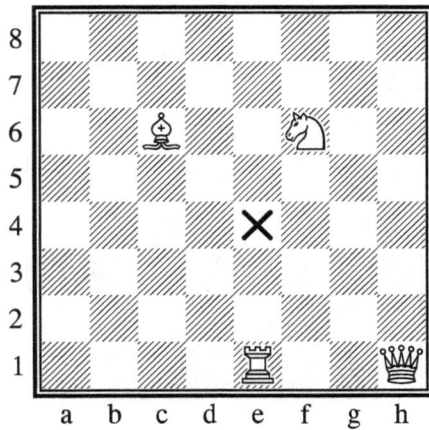

Answer _____405_____

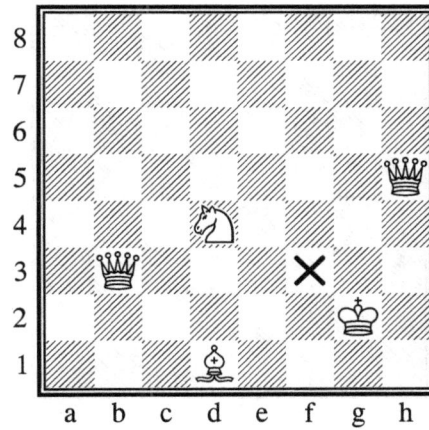

Answer _____0____

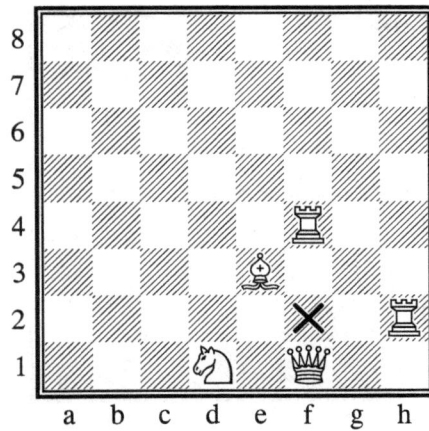

Answer _____2025_____

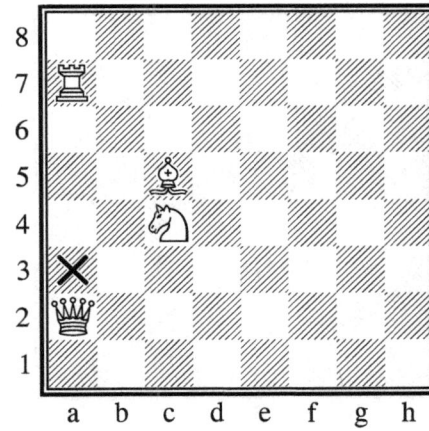

Answer _____405____

2007 - 2017 © Frank Ho, Amanda Ho, All rights reserved.　　www.homathchess.com

What is the product of all attacking pieces on the ✕ or ☒?

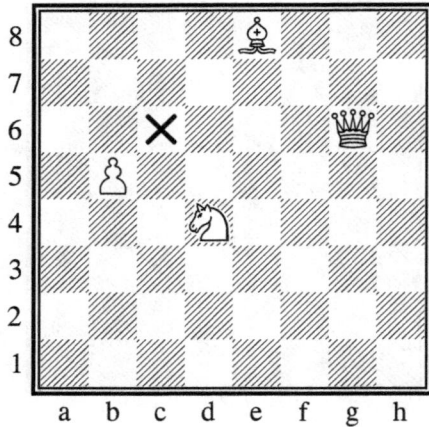

Answer _____81_____

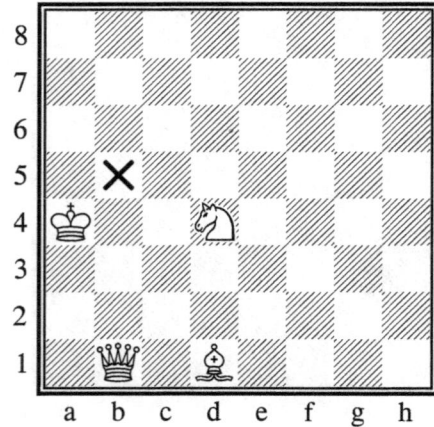

Answer _____0_____

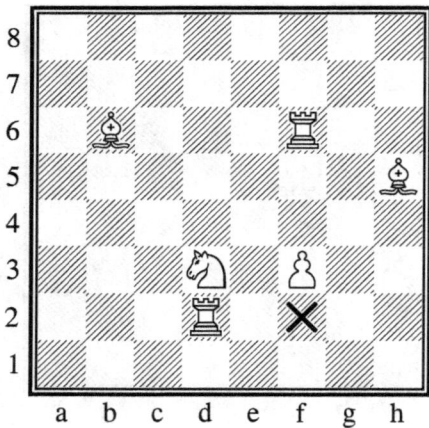

Answer _____45_____

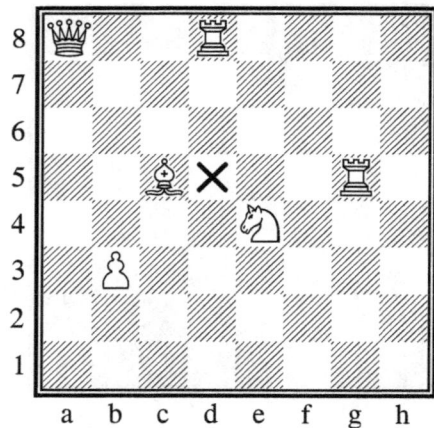

Answer _____225_____

2007 - 2017 © Frank Ho, Amanda Ho, All rights reserved.　www.homathchess.com

What is the product of all attacking pieces on the ✖ or ☒?

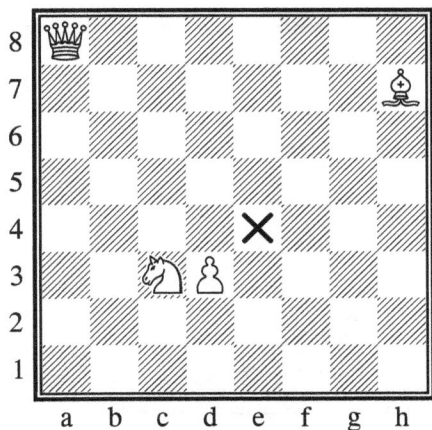

Answer _____81_____

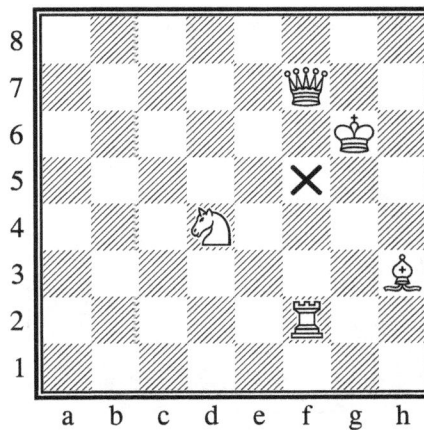

Answer _____0_____

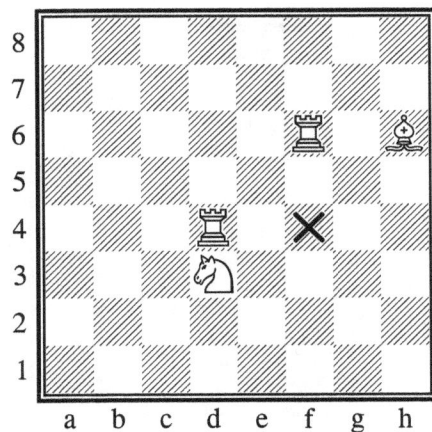

Answer _____225_____

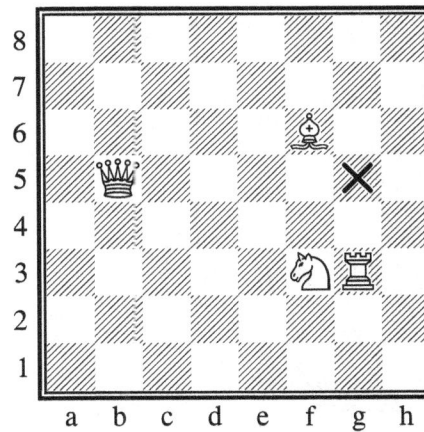

Answer _____405_____

2007 - 2017 © Frank Ho, Amanda Ho, All rights reserved.　www.homathchess.com

What is the product of all attacking pieces on the ▨ or ☒?

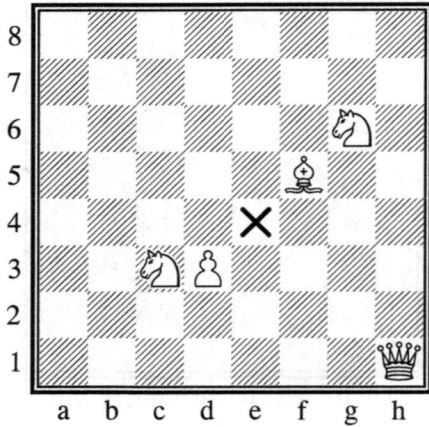

Answer _____81_____

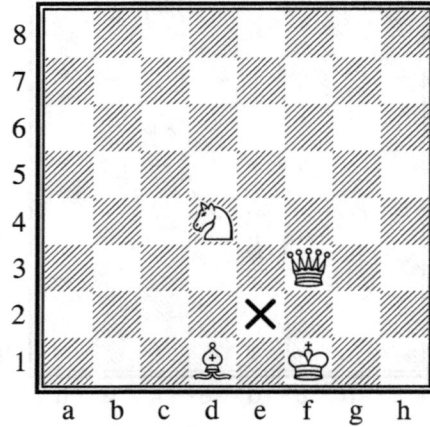

Answer _____0_____

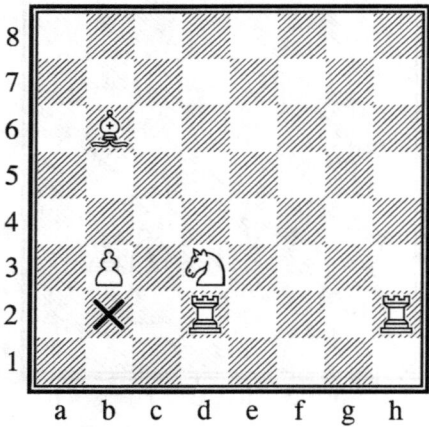

Answer _____15_____

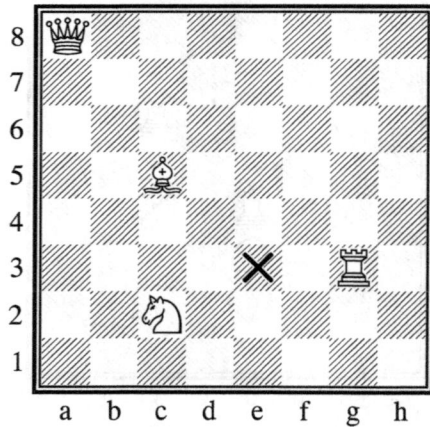

Answer _____45_____

2007 - 2017 © Frank Ho, Amanda Ho, All rights reserved.   www.homathchess.com

What is the product of all attacking pieces on the ✖ or ☒?

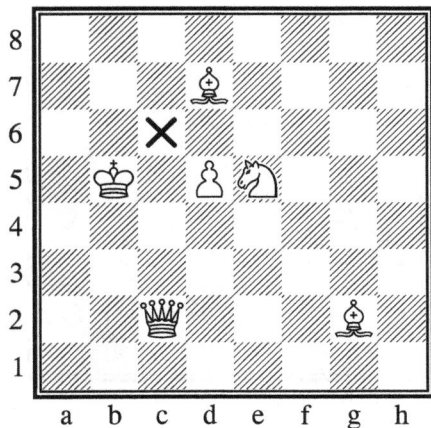

Answer _____0_____

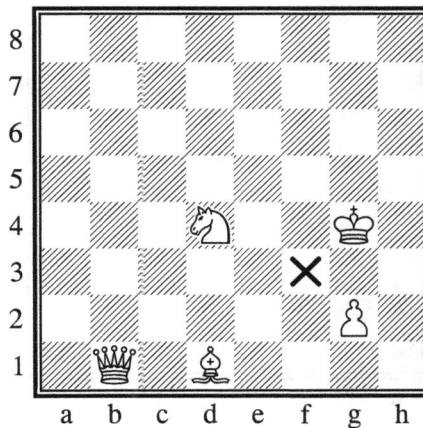

Answer _____0_____

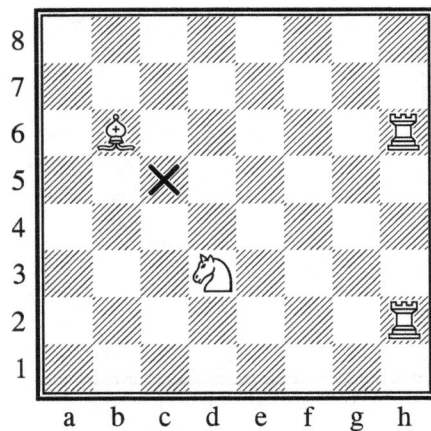

Answer _____225_____

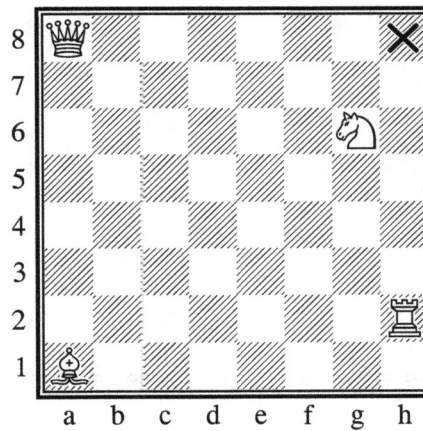

Answer _____405_____

2007 - 2017 © Frank Ho, Amanda Ho, All rights reserved.　　www.homathchess.com

What is the product of all attacking pieces on the ▨ or ☒?

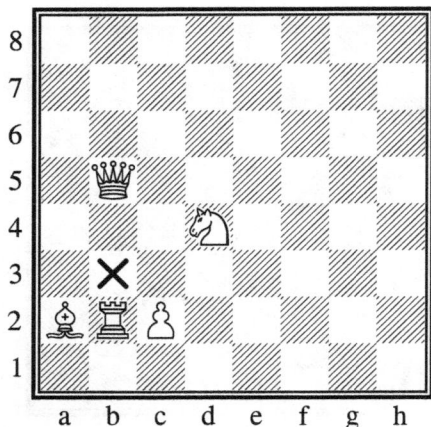

Answer _____405_____

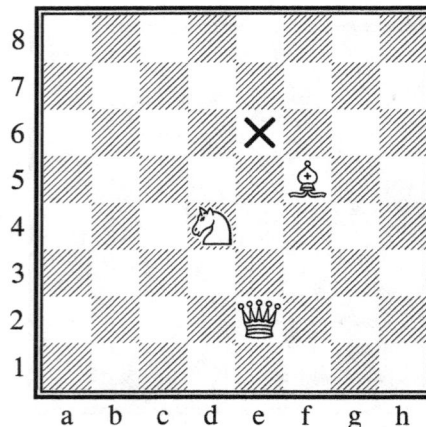

Answer _____81_____

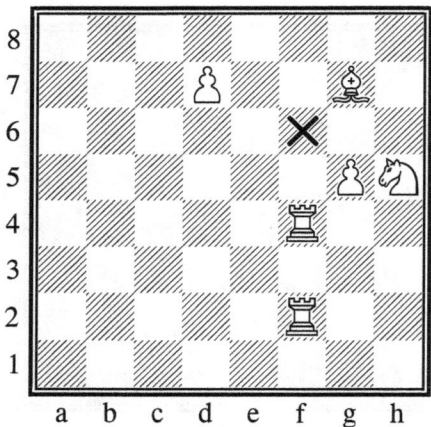

Answer _____225_____

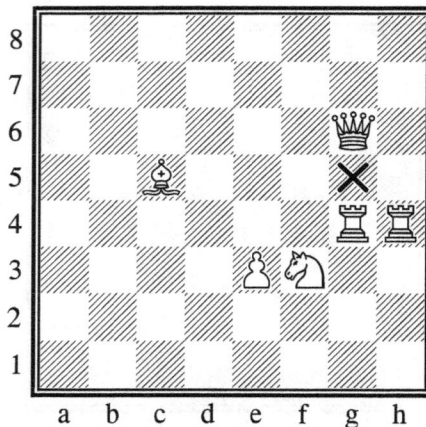

Answer _____135_____

Ho Math Chess   何数棋谜   妈！我会棋谜式乘法啦！
Mom! I Learn Multiplication Using Math-Chess-Puzzles Connection!

Student's Name _____ Date _____

2007 - 2017 © Frank Ho, Amanda Ho, All rights reserved.   www.homathchess.com

What is the product of all attacking pieces on the ▨ or ☒?

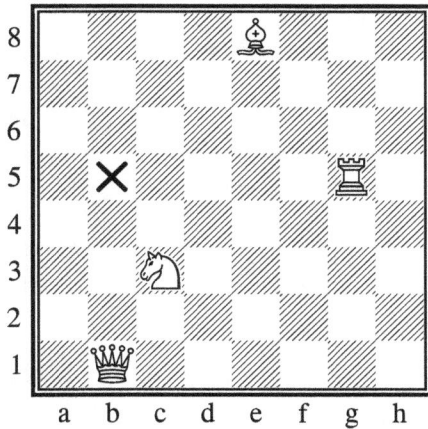

Answer _____405_____

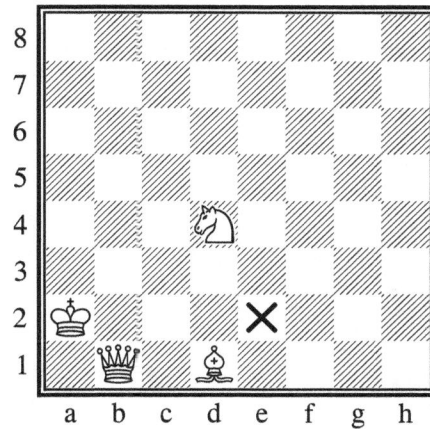

Answer _____9_____

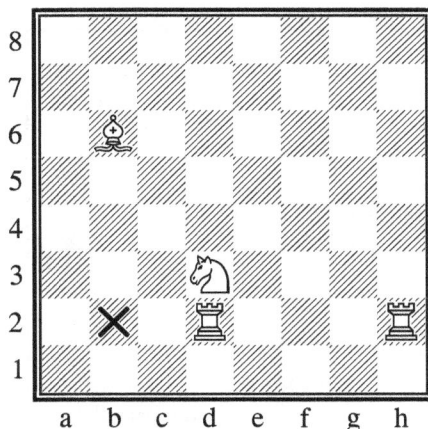

Answer _____15_____

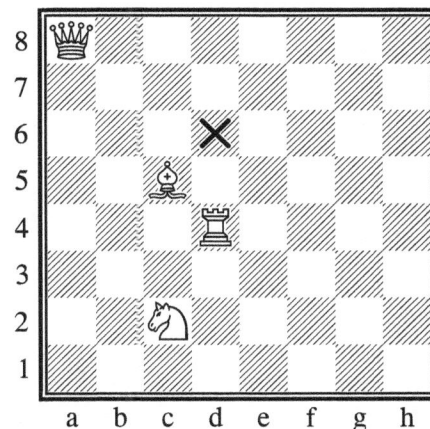

Answer _____15_____

2007 - 2017 © Frank Ho, Amanda Ho, All rights reserved.   www.homathchess.com

Mark (✗) on only the common squares controlled by all chess pieces.

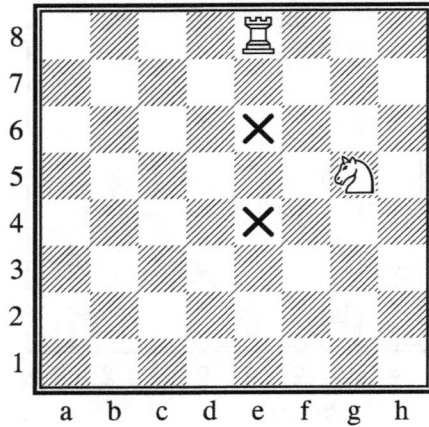

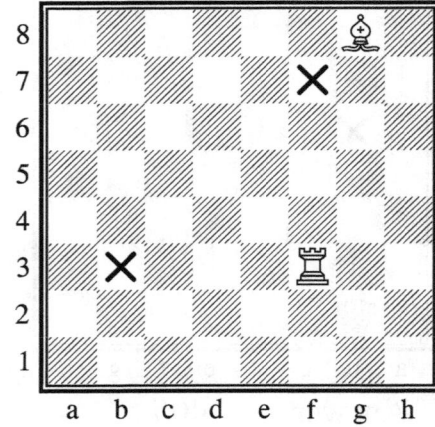

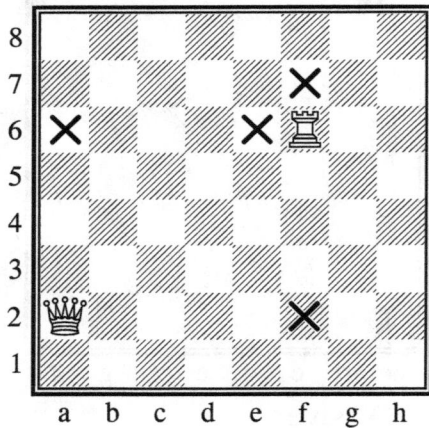

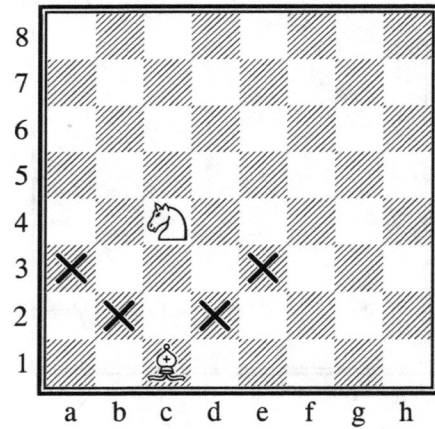

2007 - 2017 © Frank Ho, Amanda Ho, All rights reserved.   www.homathchess.com

Mark (✘) on only the common squares controlled by all chess pieces.

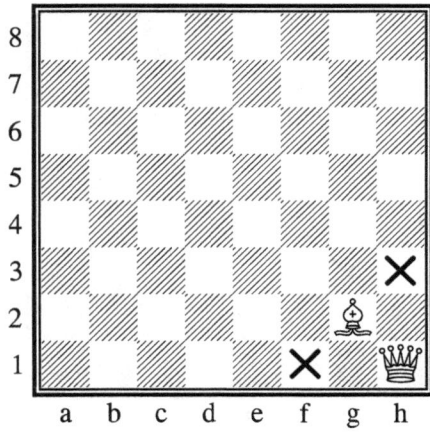

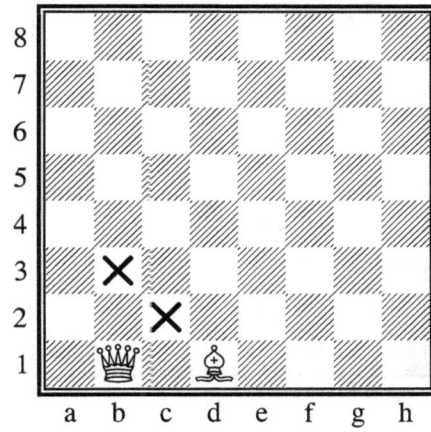

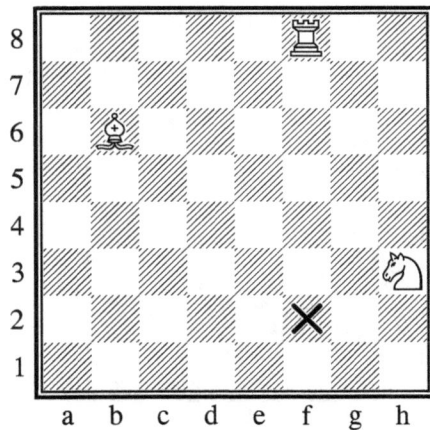

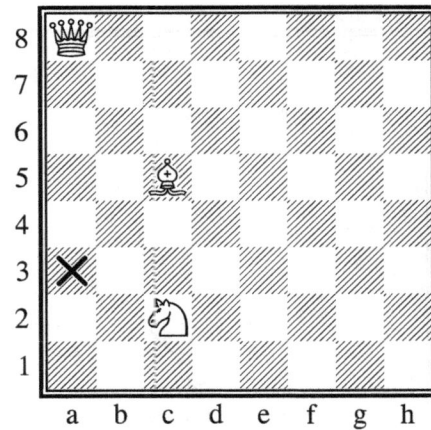

Ho Math Chess   何数棋谜   妈！我会棋谜式乘法啦！
Mom! I Learn Multiplication Using Math-Chess-Puzzles Connection!
Student's Name _____ Date _____
2007 - 2017 © Frank Ho, Amanda Ho, All rights reserved.   www.homathchess.com

Mark (✕) on only the common squares controlled by all chess pieces.

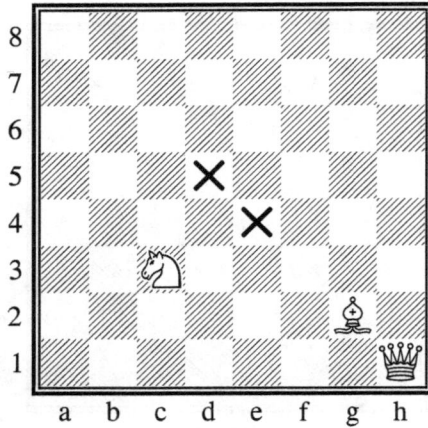

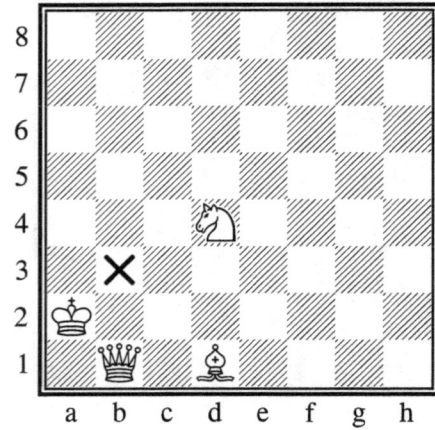

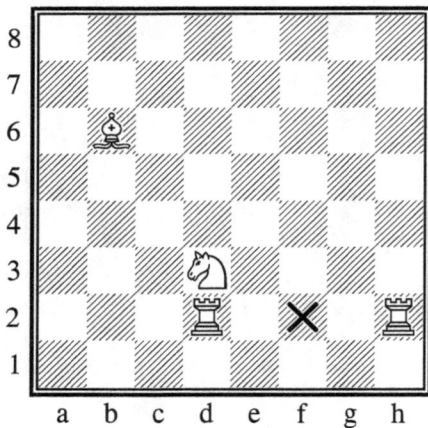

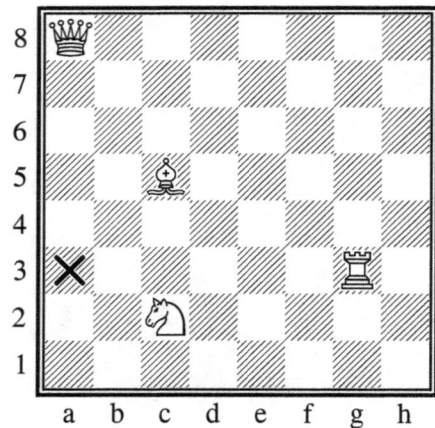

2007 - 2017 © Frank Ho, Amanda Ho, All rights reserved.   www.homathchess.com

Multiplication table

×	0	1	2	3	4	5	6	7	8	9
1	0	1	2	3	4	5	6	7	8	9
2	0	2	4	6	8	10	12	14	46	18
3	0	3	6	9	12	15	18	21	24	27
4	0	4	8	12	16	20	24	28	32	36
5	0	5	10	15	20	25	30	35	40	45
6	0	6	12	18	24	30	36	42	48	54
7	0	7	14	21	28	35	42	49	56	63
8	0	8	16	24	32	40	48	56	64	72
9	0	9	18	27	36	45	54	63	72	81

2007 - 2017 © Frank Ho, Amanda Ho, All rights reserved.   www.homathchess.com

Multiplication table

×	0	1	3	5	7	9	2	4	6	8
2	0	2	6	10	14	18	4	8	12	16
4	0	4	12	20	28	36	8	16	24	32
6	0	6	18	30	42	54	12	24	36	48
8	0	8	24	40	56	72	16	32	48	64
1	0	1	3	5	7	9	2	4	6	8
3	0	3	9	15	21	27	6	12	18	24
5	0	5	15	25	35	45	10	20	30	40
7	0	7	21	35	49	63	14	28	42	56
9	0	9	27	45	63	81	18	36	54	72

Multiplication table

Place a "/" over every multiple of 2 and "\" over every multiple of 3.

1	2	3	4	5	6	7	8	9	10
11	12	13	14	15	16	17	18	19	20
21	22	23	24	25	26	27	28	29	30
31	32	33	34	35	36	37	38	39	40
41	42	43	44	45	46	47	48	49	50
51	52	53	54	55	56	57	58	59	60
61	62	63	64	65	66	67	68	69	70
71	72	73	74	75	76	77	78	79	80
81	82	83	84	85	86	87	88	89	90
91	92	93	94	95	96	97	98	99	100

What are the unit digits of 2's multiples ? _____2, 4, 6, 8, 0_____.

The common multiples of 2 and 3 are listed as follows:
6, 12, 18, 24, 30, 36, 42, 48, 54, 60, 66, 72, 78, 84, 90, 96

_____

The common multiples of 2 and 3 are divisible by __6__.

Multiplication table

Place a "/" over every multiple of 2 and "\" over every multiple of 4.

1	2	3	4	5	6	7	8	9	10
11	12	13	14	15	16	17	18	19	20
21	22	23	24	25	26	27	28	29	30
31	32	33	34	35	36	37	38	39	40
41	42	43	44	45	46	47	48	49	50
51	52	53	54	55	56	57	58	59	60
61	62	63	64	65	66	67	68	69	70
71	72	73	74	75	76	77	78	79	80
81	82	83	84	85	86	87	88	89	90
91	92	93	94	95	96	97	98	99	100

The common multiples of 2 and 4 are listed as follows:
4, 8, 12, 16, 20, 24, 28, 32, 36, 40, 44, 48, 52, 56, 60, 64, 68, 72, 76, 80, 84, 88, 92, 96

Are the above common multiples also the 4's multiples? ____Yes_____

Ho Math Chess 何数棋谜 妈！我会棋谜式乘法啦！
Mom! I Learn Multiplication Using Math-Chess-Puzzles Connection!

Student's Name _____ Date _____

2007 - 2017 © Frank Ho, Amanda Ho, All rights reserved.   www.homathchess.com

Pattern (Preparation for equivalent fraction)

$\frac{1}{2} = \frac{2}{\square_4}$	$\frac{1}{3} = \frac{2}{\square_6}$	$\frac{1}{4} = \frac{2}{\square_8}$	$\frac{1}{5} = \frac{2}{\square_{10}}$	$\frac{1}{6} = \frac{2}{\square_{12}}$
$\frac{1}{2} = \frac{3}{\square_6}$	$\frac{1}{3} = \frac{3}{\square_9}$	$\frac{1}{4} = \frac{3}{\square_{12}}$	$\frac{1}{5} = \frac{3}{\square_{15}}$	$\frac{1}{6} = \frac{3}{\square_{18}}$
$\frac{1}{2} = \frac{4}{\square_8}$	$\frac{1}{3} = \frac{4}{\square_{12}}$	$\frac{1}{4} = \frac{4}{\square_{16}}$	$\frac{1}{5} = \frac{4}{\square_{20}}$	$\frac{1}{6} = \frac{4}{\square_{24}}$
$\frac{1}{2} = \frac{5}{\square_{10}}$	$\frac{1}{3} = \frac{5}{\square_{15}}$	$\frac{1}{4} = \frac{5}{\square_{20}}$	$\frac{1}{5} = \frac{5}{\square_{25}}$	$\frac{1}{6} = \frac{5}{\square_{30}}$
$\frac{1}{2} = \frac{6}{\square_{12}}$	$\frac{1}{3} = \frac{6}{\square_{18}}$	$\frac{1}{4} = \frac{6}{\square_{24}}$	$\frac{1}{5} = \frac{6}{\square_{30}}$	$\frac{1}{6} = \frac{6}{\square_{36}}$
$\frac{1}{2} = \frac{7}{\square_{14}}$	$\frac{1}{3} = \frac{7}{\square_{21}}$	$\frac{1}{4} = \frac{7}{\square_{28}}$	$\frac{1}{5} = \frac{7}{\square_{35}}$	$\frac{1}{6} = \frac{7}{\square_{42}}$
$\frac{1}{2} = \frac{8}{\square_{16}}$	$\frac{1}{3} = \frac{8}{\square_{24}}$	$\frac{1}{4} = \frac{8}{\square_{32}}$	$\frac{1}{5} = \frac{8}{\square_{40}}$	$\frac{1}{6} = \frac{8}{\square_{48}}$
$\frac{1}{2} = \frac{9}{\square_{18}}$	$\frac{1}{3} = \frac{9}{\square_{27}}$	$\frac{1}{4} = \frac{9}{\square_{36}}$	$\frac{1}{5} = \frac{9}{\square_{45}}$	$\frac{1}{6} = \frac{9}{\square_{54}}$

2007 - 2017 © Frank Ho, Amanda Ho, All rights reserved.   www.homathchess.com

Pattern (Preparation for equivalent fraction)

$\frac{1}{7} = \frac{2}{\Box_{14}}$	$\frac{1}{8} = \frac{2}{\Box_{16}}$	$\frac{1}{9} = \frac{2}{\Box_{18}}$	$\frac{1}{5} = \frac{2}{\Box_{10}}$	$\frac{1}{6} = \frac{2}{\Box_{12}}$
$\frac{1}{7} = \frac{3}{\Box_{21}}$	$\frac{1}{8} = \frac{3}{\Box_{24}}$	$\frac{1}{9} = \frac{3}{\Box_{27}}$	$\frac{1}{5} = \frac{3}{\Box_{15}}$	$\frac{1}{6} = \frac{3}{\Box_{18}}$
$\frac{1}{7} = \frac{4}{\Box_{28}}$	$\frac{1}{8} = \frac{4}{\Box_{32}}$	$\frac{1}{9} = \frac{4}{\Box_{36}}$	$\frac{1}{5} = \frac{4}{\Box_{20}}$	$\frac{1}{6} = \frac{4}{\Box_{24}}$
$\frac{1}{7} = \frac{5}{\Box_{35}}$	$\frac{1}{8} = \frac{5}{\Box_{40}}$	$\frac{1}{9} = \frac{5}{\Box_{45}}$	$\frac{1}{5} = \frac{5}{\Box_{25}}$	$\frac{1}{6} = \frac{5}{\Box_{30}}$
$\frac{1}{7} = \frac{6}{\Box_{42}}$	$\frac{1}{8} = \frac{6}{\Box_{48}}$	$\frac{1}{9} = \frac{6}{\Box_{54}}$	$\frac{1}{5} = \frac{6}{\Box_{30}}$	$\frac{1}{6} = \frac{6}{\Box_{36}}$
$\frac{1}{7} = \frac{7}{\Box_{49}}$	$\frac{1}{8} = \frac{7}{\Box_{56}}$	$\frac{1}{9} = \frac{7}{\Box_{63}}$	$\frac{1}{5} = \frac{7}{\Box_{35}}$	$\frac{1}{6} = \frac{7}{\Box_{42}}$
$\frac{1}{7} = \frac{8}{\Box_{56}}$	$\frac{1}{8} = \frac{8}{\Box_{64}}$	$\frac{1}{9} = \frac{8}{\Box_{72}}$	$\frac{1}{5} = \frac{8}{\Box_{40}}$	$\frac{1}{6} = \frac{8}{\Box_{48}}$
$\frac{1}{7} = \frac{9}{\Box_{63}}$	$\frac{1}{8} = \frac{9}{\Box_{72}}$	$\frac{1}{9} = \frac{9}{\Box_{81}}$	$\frac{1}{5} = \frac{9}{\Box_{45}}$	$\frac{1}{6} = \frac{9}{\Box_{54}}$

Ho Math Chess 何数棋谜 妈！我会棋谜式乘法啦！
Mom! I Learn Multiplication Using Math-Chess-Puzzles Connection!

Student's Name _____ Date _____

2007 - 2017 © Frank Ho, Amanda Ho, All rights reserved.   www.homathchess.com

Pattern (Preparation for equivalent fraction)

$\frac{1}{2} = \frac{2}{\square}4$	$\frac{1}{3} = \frac{2}{\square}6$	$\frac{1}{4} = \frac{2}{\square}8$	$\frac{1}{5} = \frac{2}{\square}10$	$\frac{1}{6} = \frac{2}{\square}12$
$\frac{1}{2} = \frac{3}{\square}6$	$\frac{1}{3} = \frac{3}{\square}9$	$\frac{1}{4} = \frac{3}{\square}12$	$\frac{1}{5} = \frac{3}{\square}15$	$\frac{1}{6} = \frac{3}{\square}18$
$\frac{1}{2} = \frac{4}{\square}8$	$\frac{1}{3} = \frac{4}{\square}12$	$\frac{1}{4} = \frac{4}{\square}16$	$\frac{1}{5} = \frac{4}{\square}20$	$\frac{1}{6} = \frac{4}{\square}24$
$\frac{1}{2} = \frac{5}{\square}10$	$\frac{1}{3} = \frac{5}{\square}15$	$\frac{1}{4} = \frac{5}{\square}20$	$\frac{1}{5} = \frac{5}{\square}25$	$\frac{1}{6} = \frac{5}{\square}30$
$\frac{1}{2} = \frac{6}{\square}12$	$\frac{1}{3} = \frac{6}{\square}18$	$\frac{1}{4} = \frac{6}{\square}24$	$\frac{1}{5} = \frac{6}{\square}30$	$\frac{1}{6} = \frac{6}{\square}36$
$\frac{1}{2} = \frac{7}{\square}14$	$\frac{1}{3} = \frac{7}{\square}21$	$\frac{1}{4} = \frac{7}{\square}28$	$\frac{1}{5} = \frac{7}{\square}35$	$\frac{1}{6} = \frac{7}{\square}42$
$\frac{1}{2} = \frac{8}{\square}16$	$\frac{1}{3} = \frac{8}{\square}24$	$\frac{1}{4} = \frac{8}{\square}32$	$\frac{1}{5} = \frac{8}{\square}40$	$\frac{1}{6} = \frac{8}{\square}48$

2007 - 2017 © Frank Ho, Amanda Ho, All rights reserved.   www.homathchess.com

Learning multiplication with multi-concept and multi-format

c	9	8	7
b	2	3	6
a	3	4	5
	1	2	3

The original square is at b2 = □.

□ × ✥ = _3_ × _8_ = _24_	□ × ✕ = 3__ × 7__ = 21__
□ × ✥ = _3 × 6_ =18_	□ × ✕ = 3__ × 5__ = 15__
□ × ✥ = _3 × 4_ = 12_	□ × ✕ = 3__ × 3__ = 9__
□ × ✥ = __3 × 2__ =6_	□ × ✕ = 3__ × 9__ =27_

$$_3\square \times _8\triangle =_{24}\bigcirc, \ _{24}\bigcirc =_8\triangle \times _3\square, \ \boxed{3}\overline{)24}^{\,\triangle}, \ \triangle\overline{)24}^{\,\boxed{3}},$$

$$\boxed{3}\,)\overline{24}^{\,\triangle}, \ \triangle\,)\overline{24}^{\,\boxed{3}},$$

$$\boxed{3}=24\times\frac{1}{\triangle}, \ \triangle=24\times\frac{1}{\boxed{3}}, \ \frac{1}{\boxed{3}}=\frac{\triangle}{24}, \ \frac{1}{\triangle}=\frac{\boxed{3}}{24}, \ \frac{24}{\triangle}=\frac{\boxed{3}}{1},$$

$$\frac{24}{\boxed{3}}=\frac{\triangle}{1}$$

2007 - 2017 © Frank Ho, Amanda Ho, All rights reserved. www.homathchess.com

Learning multiplication with multi-concept and multi-format

c	9	8	7
b	2	3	6
a	3	4	5
	1	2	3

The original square is at b2 = □.

$$_3\square \times _6\triangle =_{18}\bigcirc, \quad \bigcirc = \triangle \times \square, \quad \square \overline{)\dfrac{\triangle}{\bigcirc}}, \quad \triangle \overline{)\dfrac{\square}{\bigcirc}},$$

$$\square \overline{)\dfrac{\bigcirc}{\triangle}}, \quad \triangle \overline{)\dfrac{\bigcirc}{\square}},$$

$$\square = \bigcirc \times \dfrac{1}{\triangle}, \quad \triangle = \bigcirc \times \dfrac{1}{\square}, \quad \dfrac{1}{\square} = \dfrac{\triangle}{\bigcirc}, \quad \dfrac{1}{\triangle} = \dfrac{\square}{\bigcirc}, \quad \dfrac{\bigcirc}{\triangle} = \dfrac{\square}{1},$$

$$\dfrac{\bigcirc}{\square} = \dfrac{\triangle}{1}$$

$$_3\square \times _4\triangle =_{12}\bigcirc, \quad \bigcirc = \triangle \times \square, \quad \square \overline{)\dfrac{\triangle}{\bigcirc}}, \quad \triangle \overline{)\dfrac{\square}{\bigcirc}},$$

$$\square \overline{)\dfrac{\bigcirc}{\triangle}}, \quad \triangle \overline{)\dfrac{\bigcirc}{\square}},$$

$$\square = \bigcirc \times \dfrac{1}{\triangle}, \quad \triangle = \bigcirc \times \dfrac{1}{\square}, \quad \dfrac{1}{\square} = \dfrac{\triangle}{\bigcirc}, \quad \dfrac{1}{\square} = \dfrac{\square}{\bigcirc}, \quad \dfrac{\bigcirc}{\triangle} = \dfrac{\square}{1},$$

$$\dfrac{\bigcirc}{\square} = \dfrac{\triangle}{1}$$

2007 - 2017 © Frank Ho, Amanda Ho, All rights reserved.　　www.homathchess.com

Learning multiplication with multi-concept and multi-format

c	9	8	7
b	2	3	6
a	3	4	5
	1	2	3

The original square is at b2 = □.

$$_3\square \times {}_2\triangle = {}_6\bigcirc, \quad \bigcirc = \triangle \times \square, \quad \square\overline{)\dfrac{\triangle}{\bigcirc}}, \quad \triangle\overline{)\dfrac{\square}{\bigcirc}},$$

$$\square\overline{)\dfrac{\bigcirc}{\triangle}}, \quad \triangle\overline{)\dfrac{\bigcirc}{\square}},$$

$$\square = \bigcirc \times \dfrac{1}{\triangle}, \quad \triangle = \bigcirc \times \dfrac{1}{\square}, \quad \dfrac{1}{\square} = \dfrac{\triangle}{\bigcirc}, \quad \dfrac{1}{\triangle} = \dfrac{\square}{\bigcirc}, \quad \dfrac{\bigcirc}{\triangle} = \dfrac{\square}{1},$$

$$\dfrac{\bigcirc}{\square} = \dfrac{\triangle}{1}$$

$$_3\square \times {}_7\triangle = {}_{21}\bigcirc, \quad \bigcirc = \triangle \times \square, \quad \square\overline{)\dfrac{\triangle}{\bigcirc}}, \quad \triangle\overline{)\dfrac{\square}{\bigcirc}},$$

$$\square\overline{)\dfrac{\bigcirc}{\triangle}}, \quad \triangle\overline{)\dfrac{\bigcirc}{\square}},$$

$$\square = \bigcirc \times \dfrac{1}{\triangle}, \quad \triangle = \bigcirc \times \dfrac{1}{\square}, \quad \dfrac{1}{\square} = \dfrac{\triangle}{\bigcirc}, \quad \dfrac{1}{\triangle} = \dfrac{\square}{\bigcirc}, \quad \dfrac{\bigcirc}{\triangle} = \dfrac{\square}{1},$$

$$\dfrac{\bigcirc}{\square} = \dfrac{\triangle}{1}$$

Ho Math Chess 何数棋谜 妈！我会棋谜式乘法啦！
Mom! I Learn Multiplication Using Math-Chess-Puzzles Connection!

Student's Name _____ Date _____

2007 - 2017 © Frank Ho, Amanda Ho, All rights reserved.   www.homathchess.com

Learning multiplication with multi-concept and multi-format

c	9	8	7
b	2	3	6
a	3	4	5
	1	2	3

The original square is at b2 = □.

Ho Math Chess   何数棋谜   妈！我会棋谜式乘法啦！
Mom! I Learn Multiplication Using Math-Chess-Puzzles Connection!

Student's Name _____ Date _____

2007 - 2017 © Frank Ho, Amanda Ho, All rights reserved.   www.homathchess.com

Learning multiplication with multi-concept and multi-format

c	9	8	7
b	2	3	6
a	3	4	5
	1	2	3

The original square is at b2 = □.

Student's Name _____ Date _____

2007 - 2017 © Frank Ho, Amanda Ho, All rights reserved.   www.homathchess.com

Learning multiplication with multi-concept and multi-format

c	9	8	7
b	2	4	6
a	3	4	5
	1	2	3

The original square is at b2 = □.

□ × ⬍➤ = _4_ × _8_ = _32_

□ × ✕ = _4_ × _7=28_

□ × ⬍➤ = _4_ × _6_ = _24_

□ × ✕ = _4_ × _5=20_

□ × ⬍➤ = _4_ × _4_ = _16_

□ × ✕ = _4_ × _3=12_

□ × ⬍➤ = _4_ × _2_ = _8_

□ × ✕ = _4_ × _9= 36_

⬍➤

□ × △ = ○ , ○ = △ × □ , □⟌○̄ , △⟌○̄ , □⟌$\frac{○}{△}$,

△⟌$\frac{○}{□}$

□ = ○ × $\frac{1}{△}$, △ = ○ × $\frac{1}{□}$, $\frac{1}{□}$ = $\frac{△}{○}$, $\frac{1}{□}$ = $\frac{□}{○}$, $\frac{○}{△}$ = $\frac{□}{1}$,

$\frac{○}{□}$ = $\frac{△}{1}$

2007 - 2017 © Frank Ho, Amanda Ho, All rights reserved.   www.homathchess.com

Learning multiplication with multi-concept and multi-format

c	9	8	7
b	2	4	6
a	3	4	5
	1	2	3

The original square is at b2 = □.

$$_4\square \times _6\triangle =_{24}\bigcirc, \quad \bigcirc = \triangle \times \square, \quad \square\overline{)\dfrac{\triangle}{\bigcirc}}, \quad \triangle\overline{)\dfrac{\square}{\bigcirc}},$$

$$\square\overline{)\dfrac{\bigcirc}{\triangle}}, \quad \triangle\overline{)\dfrac{\bigcirc}{\square}},$$

$$\square = \bigcirc \times \dfrac{1}{\triangle}, \quad \triangle = \bigcirc \times \dfrac{1}{\square}, \quad \dfrac{1}{\square} = \dfrac{\triangle}{\bigcirc}, \quad \dfrac{1}{\triangle} = \dfrac{\square}{\bigcirc}, \quad \dfrac{\bigcirc}{\triangle} = \dfrac{\square}{1},$$

$$\dfrac{\bigcirc}{\square} = \dfrac{\triangle}{1}$$

$$\square_4 \times _4\triangle =_{16}\bigcirc, \quad \bigcirc = \triangle \times \square, \quad \square\overline{)\dfrac{\triangle}{\bigcirc}}, \quad \triangle\overline{)\dfrac{\square}{\bigcirc}},$$

$$\square\overline{)\dfrac{\bigcirc}{\triangle}}, \quad \triangle\overline{)\dfrac{\bigcirc}{\square}},$$

$$\square = \bigcirc \times \dfrac{1}{\triangle}, \quad \triangle = \bigcirc \times \dfrac{1}{\square}, \quad \dfrac{1}{\square} = \dfrac{\triangle}{\bigcirc}, \quad \dfrac{1}{\triangle} = \dfrac{\square}{\bigcirc}, \quad \dfrac{\bigcirc}{\triangle} = \dfrac{\square}{1},$$

$$\dfrac{\bigcirc}{\square} = \dfrac{\triangle}{1}$$

Ho Math Chess   何数棋谜   妈！我会棋谜式乘法啦！
Mom! I Learn Multiplication Using Math-Chess-Puzzles Connection!

Student's Name _____ Date _____

2007 - 2017 © Frank Ho, Amanda Ho, All rights reserved.   www.homathchess.com

Learning multiplication with multi-concept and multi-format

c	9	8	7
b	2	4	6
a	3	4	5
	1	2	3

The original square is at b2 = □.

2007 - 2017 © Frank Ho, Amanda Ho, All rights reserved.   www.homathchess.com

Learning multiplication with multi-concept and multi-format

c	9	8	7
b	2	4	6
a	3	4	5
	1	2	3

The original square is at b2 = □.

Ho Math Chess　何数棋谜　妈！我会棋谜式乘法啦！
Mom! I Learn Multiplication Using Math-Chess-Puzzles Connection!

Student's Name _____ Date _____

2007 - 2017 © Frank Ho, Amanda Ho, All rights reserved.　www.homathchess.com

Learning multiplication with multi-concept and multi-format

c	9	8	7
b	2	4	6
a	3	4	5
	1	2	3

The original square is at b2 = □.

$$\square_4 \times {}_9\triangle = {}_{36}\bigcirc, \quad \bigcirc = \triangle \times \square, \quad \square\overline{)\dfrac{\triangle}{\bigcirc}}, \quad \triangle\overline{)\dfrac{\square}{\bigcirc}},$$

$$\square\overline{)\dfrac{\bigcirc}{\triangle}}, \quad \triangle\overline{)\dfrac{\bigcirc}{\square}},$$

$$\square = \bigcirc \times \dfrac{1}{\triangle}, \quad \triangle = \bigcirc \times \dfrac{1}{\square}, \quad \dfrac{1}{\square} = \dfrac{\triangle}{\bigcirc}, \quad \dfrac{1}{\triangle} = \dfrac{\square}{\bigcirc}, \quad \dfrac{\bigcirc}{\triangle} = \dfrac{\square}{1},$$

$$\dfrac{\bigcirc}{\square} = \dfrac{\triangle}{1}$$

2007 - 2017 © Frank Ho, Amanda Ho, All rights reserved.　　www.homathchess.com

Learning multiplication with multi-concept and multi-format

c	9	8	7
b	2	5	6
a	3	4	5
	1	2	3

The original square is at b2 = □.

□ × ⬌⬍ = _5_ × _8_ = _40_

□ × ⬌⬍ = _5_ × 6 = 30

□ × ⬌⬍ = _5_ × 4 = 20

□ × ⬌⬍ = _5_ × 2 = 10

□ × ✕ = _5_ × _7_ = _35_

□ × ✕ = _5_ × 5 = 25

□ × ✕ = _5_ × 3 = 15

□ × ✕ = _5_ × 9 = 45

$$\Box \times \triangle = \bigcirc,\ \bigcirc = \triangle \times \Box,\ \Box\overline{)\bigcirc},\ \triangle\overline{)\bigcirc},\ \dfrac{\Box}{\triangle}\bigcirc,\ \triangle\dfrac{)\bigcirc}{\Box},$$

$$\Box = \bigcirc \times \dfrac{1}{\triangle},\ \triangle = \bigcirc \times \dfrac{1}{\Box},\ \dfrac{1}{\Box} = \dfrac{\triangle}{\bigcirc},\ \dfrac{1}{\triangle} = \dfrac{\Box}{\bigcirc},\ \dfrac{\bigcirc}{\triangle} = \dfrac{\Box}{1},$$

$$\dfrac{\bigcirc}{\Box} = \dfrac{\triangle}{1}$$

Ho Math Chess   何数棋谜   妈！我会棋谜式乘法啦！
Mom! I Learn Multiplication Using Math-Chess-Puzzles Connection!

Student's Name _____ Date _____

2007 - 2017 © Frank Ho, Amanda Ho, All rights reserved.   www.homathchess.com

Learning multiplication with multi-concept and multi-format

c	9	8	7
b	2	5	6
a	3	4	5
	1	2	3

The original square is at b2 = □.

$$\square \times \triangle = \bigcirc, \quad \bigcirc = \triangle \times \square, \quad \square\overline{)\dfrac{\triangle}{\bigcirc}}, \quad \triangle\overline{)\dfrac{\square}{\bigcirc}}, \quad \dfrac{\square}{\triangle}\dfrac{\bigcirc}{}, \quad \triangle\dfrac{\bigcirc}{\square},$$

$$\square = \bigcirc \times \dfrac{1}{\triangle}, \quad \triangle = \bigcirc \times \dfrac{1}{\square}, \quad \dfrac{1}{\square} = \dfrac{\triangle}{\bigcirc}, \quad \dfrac{1}{\square} = \dfrac{\square}{\bigcirc}, \quad \dfrac{\bigcirc}{\triangle} = \dfrac{\square}{1},$$

$$\dfrac{\bigcirc}{\square} = \dfrac{\triangle}{1}$$

$$\square \times \triangle = \bigcirc, \quad \bigcirc = \triangle \times \square, \quad \square\overline{)\dfrac{\triangle}{\bigcirc}}, \quad \triangle\overline{)\dfrac{\square}{\bigcirc}}, \quad \dfrac{\square}{\triangle}\dfrac{\bigcirc}{}, \quad \triangle\dfrac{\bigcirc}{\square},$$

$$\square = \bigcirc \times \dfrac{1}{\triangle}, \quad \triangle = \bigcirc \times \dfrac{1}{\square}, \quad \dfrac{1}{\square} = \dfrac{\triangle}{\bigcirc}, \quad \dfrac{1}{\square} = \dfrac{\square}{\bigcirc}, \quad \dfrac{\bigcirc}{\triangle} = \dfrac{\square}{1},$$

$$\dfrac{\bigcirc}{\square} = \dfrac{\triangle}{1}$$

2007 - 2017 © Frank Ho, Amanda Ho, All rights reserved.　　www.homathchess.com

Learning multiplication with multi-concept and multi-format

c	9	8	7
b	2	5	6
a	3	4	5
	1	2	3

The original square is at b2 = □.

$\square \times \triangle = \bigcirc$, $\bigcirc = \triangle \times \square$, $\square \overline{)\cfrac{\triangle}{\bigcirc}}$, $\triangle \overline{)\cfrac{\square}{\bigcirc}}$, $\cfrac{\square \cdot \bigcirc}{\triangle}$, $\triangle \cdot \cfrac{\bigcirc}{\square}$

$\square = \bigcirc \times \cfrac{1}{\triangle}$, $\triangle = \bigcirc \times \cfrac{1}{\square}$, $\cfrac{1}{\square} = \cfrac{\triangle}{\bigcirc}$, $\cfrac{1}{\triangle} = \cfrac{\square}{\bigcirc}$, $\cfrac{\bigcirc}{\triangle} = \cfrac{\square}{1}$, $\cfrac{\bigcirc}{\square} = \cfrac{\triangle}{1}$

$\square \times \triangle = \bigcirc$, $\bigcirc = \triangle \times \square$, $\square \overline{)\cfrac{\triangle}{\bigcirc}}$, $\triangle \overline{)\cfrac{\square}{\bigcirc}}$, $\cfrac{\square \cdot \bigcirc}{\triangle}$, $\triangle \cdot \cfrac{\bigcirc}{\square}$

$\square = \bigcirc \times \cfrac{1}{\triangle}$, $\triangle = \bigcirc \times \cfrac{1}{\square}$, $\cfrac{1}{\square} = \cfrac{\triangle}{\bigcirc}$, $\cfrac{1}{\triangle} = \cfrac{\square}{\bigcirc}$, $\cfrac{\bigcirc}{\triangle} = \cfrac{\square}{1}$, $\cfrac{\bigcirc}{\square} = \cfrac{\triangle}{1}$

2007 - 2017 © Frank Ho, Amanda Ho, All rights reserved.　www.homathchess.com

Learning multiplication with multi-concept and multi-format

c	9	8	7
b	2	5	6
a	3	4	5
	1	2	3

The original square is at b2 = □.

$$\square \times \triangle = \bigcirc, \quad \bigcirc = \triangle \times \square, \quad \square \overline{)\dfrac{\triangle}{\bigcirc}}, \quad \triangle \overline{)\dfrac{\square}{\bigcirc}}, \quad \square \dfrac{\bigcirc}{\triangle},$$

$$\triangle \dfrac{\bigcirc}{\square}$$

$$\square = \bigcirc \times \dfrac{1}{\triangle}, \quad \triangle = \bigcirc \times \dfrac{1}{\square}, \quad \dfrac{1}{\square} = \dfrac{\triangle}{\bigcirc}, \quad \dfrac{1}{\triangle} = \dfrac{\square}{\bigcirc}, \quad \dfrac{\bigcirc}{\triangle} = \dfrac{\square}{1},$$

$$\dfrac{\bigcirc}{\square} = \dfrac{\triangle}{1}$$

$$\square \times \triangle = \bigcirc, \quad \bigcirc = \triangle \times \square, \quad \square \overline{)\dfrac{\triangle}{\bigcirc}}, \quad \triangle \overline{)\dfrac{\square}{\bigcirc}}, \quad \square \dfrac{\bigcirc}{\triangle},$$

$$\triangle \dfrac{\bigcirc}{\square}$$

$$\square = \bigcirc \times \dfrac{1}{\triangle}, \quad \triangle = \bigcirc \times \dfrac{1}{\square}, \quad \dfrac{1}{\square} = \dfrac{\triangle}{\bigcirc}, \quad \dfrac{1}{\triangle} = \dfrac{\square}{\bigcirc}, \quad \dfrac{\bigcirc}{\triangle} = \dfrac{\square}{1},$$

$$\dfrac{\bigcirc}{\square} = \dfrac{\triangle}{1}$$

Ho Math Chess 何数棋谜 妈！我会棋谜式乘法啦！
Mom! I Learn Multiplication Using Math-Chess-Puzzles Connection!

Student's Name _____ Date _____

2007 - 2017 © Frank Ho, Amanda Ho, All rights reserved.   www.homathchess.com

Learning multiplication with multi-concept and multi-format

c	9	8	7
b	2	5	6
a	3	4	5
	1	2	3

The original square is at b2 = □.

$$\square \times \triangle = \bigcirc, \quad \bigcirc = \triangle \times \square, \quad \square\overline{)\bigcirc}, \quad \triangle\overline{)\bigcirc}, \quad \square\frac{\bigcirc}{\triangle},$$

$$\triangle\frac{\bigcirc}{\square}$$

$$\square = \bigcirc \times \frac{1}{\triangle}, \quad \triangle = \bigcirc \times \frac{1}{\square}, \quad \frac{1}{\square} = \frac{\triangle}{\bigcirc}, \quad \frac{1}{\triangle} = \frac{\square}{\bigcirc}, \quad \frac{\bigcirc}{\triangle} = \frac{\square}{1},$$

$$\frac{\bigcirc}{\square} = \frac{\triangle}{1}$$

Student's Name _____ Date _____

2007 - 2017 © Frank Ho, Amanda Ho, All rights reserved.   www.homathchess.com

Learning multiplication with multi-concept and multi-format

c	9	8	7
b	2	6	6
a	3	4	5
	1	2	3

The original square is at b2 = □.

□ × ⬌ = _6_ ×8_ =48__	□ × ✗ = _6_×7__ =42__
□ × ⬌ = _6_×6_ = _36_	□ × ✗ = _6_×5__ = 30__
□ × ⬌ = _6_×4_ =24	□ × ✗ = _6_×3__ = 18__
□ × ⬌ = _6_×2__ = 12__	□ × ✗ = _6_×9__ =54__

⬌

$$\square \times \triangle = \bigcirc, \quad \bigcirc = \triangle \times \square, \quad \square\overline{)\bigcirc}, \quad \triangle\overline{)\bigcirc}, \quad \square\frac{\bigcirc}{\triangle},$$

$$\triangle\frac{\bigcirc}{\square}$$

$$\square = \bigcirc \times \frac{1}{\triangle}, \quad \triangle = \bigcirc \times \frac{1}{\square}, \quad \frac{1}{\square} = \frac{\triangle}{\bigcirc}, \quad \frac{1}{\square} = \frac{\square}{\bigcirc}, \quad \frac{\bigcirc}{\triangle} = \frac{\square}{1},$$

$$\frac{\bigcirc}{\square} = \frac{\triangle}{1}$$

Mom! I Learn Multiplication Using Math-Chess-Puzzles Connection!

Student's Name _____ Date _____

2007 - 2017 © Frank Ho, Amanda Ho, All rights reserved.　　www.homathchess.com

Learning multiplication with multi-concept and multi-format

c	9	8	7
b	2	6	6
a	3	4	5
	1	2	3

The original square is at b2 = □.

Ho Math Chess 何数棋谜 妈！我会棋谜式乘法啦！
Mom! I Learn Multiplication Using Math-Chess-Puzzles Connection!

Student's Name _____ Date _____

2007 - 2017 © Frank Ho, Amanda Ho, All rights reserved. www.homathchess.com

Learning multiplication with multi-concept and multi-format

c	9	8	7
b	2	6	6
a	3	4	5
	1	2	3

The original square is at b2 = □.

$$\square \times \triangle = \bigcirc, \quad \bigcirc = \triangle \times \square, \quad \square\overline{)\dfrac{\triangle}{\bigcirc}}, \quad \triangle\overline{)\dfrac{\square}{\bigcirc}}, \quad \square\dfrac{\bigcirc}{\triangle},$$

$$\triangle\dfrac{\bigcirc}{\square}$$

$$\square = \bigcirc \times \dfrac{1}{\triangle}, \quad \triangle = \bigcirc \times \dfrac{1}{\square}, \quad \dfrac{1}{\square} = \dfrac{\triangle}{\bigcirc}, \quad \dfrac{1}{\triangle} = \dfrac{\square}{\bigcirc}, \quad \dfrac{\bigcirc}{\triangle} = \dfrac{\square}{1},$$

$$\dfrac{\bigcirc}{\square} = \dfrac{\triangle}{1}$$

$$\square \times \triangle = \bigcirc, \quad \bigcirc = \triangle \times \square, \quad \square\overline{)\dfrac{\triangle}{\bigcirc}}, \quad \triangle\overline{)\dfrac{\square}{\bigcirc}}, \quad \square\dfrac{\bigcirc}{\triangle},$$

$$\triangle\dfrac{\bigcirc}{\square}$$

$$\square = \bigcirc \times \dfrac{1}{\triangle}, \quad \triangle = \bigcirc \times \dfrac{1}{\square}, \quad \dfrac{1}{\square} = \dfrac{\triangle}{\bigcirc}, \quad \dfrac{1}{\triangle} = \dfrac{\square}{\bigcirc}, \quad \dfrac{\bigcirc}{\triangle} = \dfrac{\square}{1},$$

$$\dfrac{\bigcirc}{\square} = \dfrac{\triangle}{1}$$

2007 - 2017 © Frank Ho, Amanda Ho, All rights reserved.   www.homathchess.com

Learning multiplication with multi-concept and multi-format

c	9	8	7
b	2	6	6
a	3	4	5
	1	2	3

The original square is at b2 = □.

Ho Math Chess 何数棋谜 妈！我会棋谜式乘法啦！
Mom! I Learn Multiplication Using Math-Chess-Puzzles Connection!

Student's Name _____ Date _____

2007 - 2017 © Frank Ho, Amanda Ho, All rights reserved.   www.homathchess.com

Learning multiplication with multi-concept and multi-format

c	9	8	7
b	2	6	6
a	3	4	5
	1	2	3

The original square is at b2 = □.

$$\square \times \triangle = \bigcirc, \quad \bigcirc = \triangle \times \square, \quad \square \overline{)\dfrac{\triangle}{\bigcirc}}, \quad \triangle \overline{)\dfrac{\square}{\bigcirc}}, \quad \dfrac{\square}{} \overline{)\dfrac{\bigcirc}{\triangle}},$$

$$\triangle \overline{)\dfrac{\bigcirc}{\square}}$$

$$\square = \bigcirc \times \dfrac{1}{\triangle}, \quad \triangle = \bigcirc \times \dfrac{1}{\square}, \quad \dfrac{1}{\square} = \dfrac{\triangle}{\bigcirc}, \quad \dfrac{1}{\triangle} = \dfrac{\square}{\bigcirc}, \quad \dfrac{\bigcirc}{\triangle} = \dfrac{\square}{1},$$

$$\dfrac{\bigcirc}{\square} = \dfrac{\triangle}{1}$$

Ho Math Chess 何数棋谜 妈!我会棋谜式乘法啦!
Mom! I Learn Multiplication Using Math-Chess-Puzzles Connection!

Student's Name _____ Date _____

2007 - 2017 © Frank Ho, Amanda Ho, All rights reserved.   www.homathchess.com

Learning multiplication with multi-concept and multi-format

c	9	8	7
b	2	7	6
a	3	4	5
	1	2	3

The original square is at b2 = ☐.

☐ × ⬍ = $_7 × 8$ = 56	☐ × ✕ = $_7 × 7$ = 49
☐ × ⬌ = $_7 × 6$ = 42	☐ × ✕ = $_7 × 5$ = 35
☐ × ⬍ = $_7 × 4$ = 28	☐ × ✕ = $_7 × 3$ = 21
☐ × ⬍ = $_7 × 2$ = 14	☐ × ✕ = $_7 × 9$ = 63

⬌

$$\square \times \triangle = \bigcirc, \quad \bigcirc = \triangle \times \square, \quad \square\overline{)\bigcirc}, \quad \triangle\overline{)\bigcirc}, \quad \dfrac{\square}{\triangle}, $$

$$\triangle\overline{)\dfrac{\bigcirc}{\square}}$$

$$\square = \bigcirc \times \dfrac{1}{\triangle}, \quad \triangle = \bigcirc \times \dfrac{1}{\square}, \quad \dfrac{1}{\square} = \dfrac{\triangle}{\bigcirc}, \quad \dfrac{1}{\square} = \dfrac{\square}{\bigcirc}, \quad \dfrac{\bigcirc}{\triangle} = \dfrac{\square}{1},$$

$$\dfrac{\bigcirc}{\square} = \dfrac{\triangle}{1}$$

254

2007 - 2017 © Frank Ho, Amanda Ho, All rights reserved. www.homathchess.com

Learning multiplication with multi-concept and multi-format

c	9	8	7
b	2	7	6
a	3	4	5
	1	2	3

The original square is at b2 = □.

$$\square \times \triangle = \bigcirc, \ \bigcirc = \triangle \times \square, \ \square \overline{)\dfrac{\triangle}{\bigcirc}}, \ \triangle \overline{)\dfrac{\square}{\bigcirc}}, \ \square \overline{)\dfrac{\bigcirc}{\triangle}},$$

$$\triangle \overline{)\dfrac{\bigcirc}{\square}}$$

$$\square = \bigcirc \times \dfrac{1}{\triangle}, \quad \triangle = \bigcirc \times \dfrac{1}{\square}, \quad \dfrac{1}{\square} = \dfrac{\triangle}{\bigcirc}, \quad \dfrac{1}{\triangle} = \dfrac{\square}{\bigcirc}, \quad \dfrac{\bigcirc}{\triangle} = \dfrac{\square}{1},$$

$$\dfrac{\bigcirc}{\square} = \dfrac{\triangle}{1}$$

$$\square \times \triangle = \bigcirc, \ \bigcirc = \triangle \times \square, \ \square \overline{)\dfrac{\triangle}{\bigcirc}}, \ \triangle \overline{)\dfrac{\square}{\bigcirc}}, \ \square \overline{)\dfrac{\bigcirc}{\triangle}},$$

$$\triangle \overline{)\dfrac{\bigcirc}{\square}}$$

$$\square = \bigcirc \times \dfrac{1}{\triangle}, \quad \triangle = \bigcirc \times \dfrac{1}{\square}, \quad \dfrac{1}{\square} = \dfrac{\triangle}{\bigcirc}, \quad \dfrac{1}{\triangle} = \dfrac{\square}{\bigcirc}, \quad \dfrac{\bigcirc}{\triangle} = \dfrac{\square}{1},$$

$$\dfrac{\bigcirc}{\square} = \dfrac{\triangle}{1}$$

Ho Math Chess   何数棋谜   妈！我会棋谜式乘法啦！
Mom! I Learn Multiplication Using Math-Chess-Puzzles Connection!

Student's Name _____ Date _____

2007 - 2017 © Frank Ho, Amanda Ho, All rights reserved.   www.homathchess.com

Learning multiplication with multi-concept and multi-format

c	9	8	7
b	2	7	6
a	3	4	5
	1	2	3

The original square is at b2 = □.

$$\square \times \triangle = \bigcirc, \quad \bigcirc = \triangle \times \square, \quad \square\overline{)\dfrac{\triangle}{\bigcirc}}, \quad \triangle\overline{)\dfrac{\square}{\bigcirc}}, \quad \square\dfrac{\bigcirc}{\triangle}, \quad \triangle\dfrac{\bigcirc}{\square}$$

$$\square = \bigcirc \times \dfrac{1}{\triangle}, \quad \triangle = \bigcirc \times \dfrac{1}{\square}, \quad \dfrac{1}{\square} = \dfrac{\triangle}{\bigcirc}, \quad \dfrac{1}{\triangle} = \dfrac{\square}{\bigcirc}, \quad \dfrac{\bigcirc}{\triangle} = \dfrac{\square}{1}, \quad \dfrac{\bigcirc}{\square} = \dfrac{\triangle}{1}$$

$$\square \times \triangle = \bigcirc, \quad \bigcirc = \triangle \times \square, \quad \square\overline{)\dfrac{\triangle}{\bigcirc}}, \quad \triangle\overline{)\dfrac{\square}{\bigcirc}}, \quad \square\dfrac{\bigcirc}{\triangle}, \quad \triangle\dfrac{\bigcirc}{\square}$$

$$\square = \bigcirc \times \dfrac{1}{\triangle}, \quad \triangle = \bigcirc \times \dfrac{1}{\square}, \quad \dfrac{1}{\square} = \dfrac{\triangle}{\bigcirc}, \quad \dfrac{1}{\triangle} = \dfrac{\square}{\bigcirc}, \quad \dfrac{\bigcirc}{\triangle} = \dfrac{\square}{1}, \quad \dfrac{\bigcirc}{\square} = \dfrac{\triangle}{1}$$

2007 - 2017 © Frank Ho, Amanda Ho, All rights reserved.   www.homathchess.com

Learning multiplication with multi-concept and multi-format

c	9	8	7
b	2	7	6
a	3	4	5
	1	2	3

The original square is at b2 = □.

$$\square \times \triangle = \bigcirc, \quad \bigcirc = \triangle \times \square, \quad \square\overline{)\dfrac{\triangle}{\bigcirc}}, \quad \triangle\overline{)\dfrac{\square}{\bigcirc}}, \quad \square\dfrac{\bigcirc}{\triangle},$$

$$\triangle\dfrac{\bigcirc}{\square}$$

$$\square = \bigcirc \times \dfrac{1}{\triangle}, \quad \triangle = \bigcirc \times \dfrac{1}{\square}, \quad \dfrac{1}{\square} = \dfrac{\triangle}{\bigcirc}, \quad \dfrac{1}{\triangle} = \dfrac{\square}{\bigcirc}, \quad \dfrac{\bigcirc}{\triangle} = \dfrac{\square}{1},$$

$$\dfrac{\bigcirc}{\square} = \dfrac{\triangle}{1}$$

$$\square \times \triangle = \bigcirc, \quad \bigcirc = \triangle \times \square, \quad \square\overline{)\dfrac{\triangle}{\bigcirc}}, \quad \triangle\overline{)\dfrac{\square}{\bigcirc}}, \quad \square\dfrac{\bigcirc}{\triangle},$$

$$\triangle\dfrac{\bigcirc}{\square}$$

$$\square = \bigcirc \times \dfrac{1}{\triangle}, \quad \triangle = \bigcirc \times \dfrac{1}{\square}, \quad \dfrac{1}{\square} = \dfrac{\triangle}{\bigcirc}, \quad \dfrac{1}{\triangle} = \dfrac{\square}{\bigcirc}, \quad \dfrac{\bigcirc}{\triangle} = \dfrac{\square}{1},$$

$$\dfrac{\bigcirc}{\square} = \dfrac{\triangle}{1}$$

Ho Math Chess   何数棋谜   妈！我会棋谜式乘法啦！

Mom! I Learn Multiplication Using Math-Chess-Puzzles Connection!

Student's Name _____ Date _____

2007 - 2017 © Frank Ho, Amanda Ho, All rights reserved.   www.homathchess.com

Learning multiplication with multi-concept and multi-format

c	9	8	7
b	2	7	6
a	3	4	5
	1	2	3

The original square is at b2 = □.

Ho Math Chess　何数棋谜　妈！我会棋谜式乘法啦！
Mom! I Learn Multiplication Using Math-Chess-Puzzles Connection!

Student's Name _____　Date _____

2007 - 2017 © Frank Ho, Amanda Ho, All rights reserved.　　www.homathchess.com

Learning multiplication with multi-concept and multi-format

c	9	8	7
b	2	8	6
a	3	4	5
	1	2	3

The original square is at b2 = □.

$\square \times$ ✥ $= \_8 \times 8\_ = 64\_$

$\square \times$ ✕ $= \_8 \times 7\_ = 56\_$

$\square \times$ ✥ $= \_8 \times 6\_ = 48\_$

$\square \times$ ✕ $= \_8 \times 5\_ = 40\_$

$\square \times$ ✥ $= \_8 \times 4\_ = 32\_$

$\square \times$ ✕ $= \_8 \times 3\_ = 24\_$

$\square \times$ ✥ $= \_8 \times 2\_ = 16\_$

$\square \times$ ✕ $= \_8 \times 9\_ = 72\_$

✥

$\square \times \triangle = \bigcirc$, $\bigcirc = \triangle \times \square$, $\square \overline{)\dfrac{\triangle}{\bigcirc}}$, $\triangle \overline{)\dfrac{\square}{\bigcirc}}$, $\dfrac{\square \overline{)\bigcirc}}{\triangle}$, $\triangle \overline{)\dfrac{\bigcirc}{\square}}$

$\square = \bigcirc \times \dfrac{1}{\triangle}$, $\triangle = \bigcirc \times \dfrac{1}{\square}$, $\dfrac{1}{\square} = \dfrac{\triangle}{\bigcirc}$, $\dfrac{1}{\triangle} = \dfrac{\square}{\bigcirc}$, $\dfrac{\bigcirc}{\triangle} = \dfrac{\square}{1}$,

$\dfrac{\bigcirc}{\square} = \dfrac{\triangle}{1}$

Ho Math Chess 何数棋谜 妈！我会棋谜式乘法啦！
Mom! I Learn Multiplication Using Math-Chess-Puzzles Connection!

Student's Name _____ Date _____

2007 - 2017 © Frank Ho, Amanda Ho, All rights reserved.   www.homathchess.com

Learning multiplication with multi-concept and multi-format

c	9	8	7
b	2	8	6
a	3	4	5
	1	2	3

The original square is at b2 = □.

$$\square \times \triangle = \bigcirc, \quad \bigcirc = \triangle \times \square, \quad \square\overline{)\dfrac{\triangle}{\bigcirc}}, \quad \triangle\overline{)\dfrac{\square}{\bigcirc}}, \quad \dfrac{\square}{\triangle}\bigcirc,$$

$$\triangle\overline{)\dfrac{\bigcirc}{\square}}$$

$$\square = \bigcirc \times \dfrac{1}{\triangle}, \quad \triangle = \bigcirc \times \dfrac{1}{\square}, \quad \dfrac{1}{\square} = \dfrac{\triangle}{\bigcirc}, \quad \dfrac{1}{\triangle} = \dfrac{\square}{\bigcirc}, \quad \dfrac{\bigcirc}{\triangle} = \dfrac{\square}{1},$$

$$\dfrac{\bigcirc}{\square} = \dfrac{\triangle}{1}$$

$$\square \times \triangle = \bigcirc, \quad \bigcirc = \triangle \times \square, \quad \square\overline{)\dfrac{\triangle}{\bigcirc}}, \quad \triangle\overline{)\dfrac{\square}{\bigcirc}}, \quad \dfrac{\square}{\triangle}\bigcirc,$$

$$\triangle\overline{)\dfrac{\bigcirc}{\square}}$$

$$\square = \bigcirc \times \dfrac{1}{\triangle}, \quad \triangle = \bigcirc \times \dfrac{1}{\square}, \quad \dfrac{1}{\square} = \dfrac{\triangle}{\bigcirc}, \quad \dfrac{1}{\triangle} = \dfrac{\square}{\bigcirc}, \quad \dfrac{\bigcirc}{\triangle} = \dfrac{\square}{1},$$

$$\dfrac{\bigcirc}{\square} = \dfrac{\triangle}{1}$$

2007 - 2017 © Frank Ho, Amanda Ho, All rights reserved.   www.homathchess.com

Learning multiplication with multi-concept and multi-format

c	9	8	7
b	2	8	6
a	3	4	5
	1	2	3

The original square is at b2 = □.

$$\square \times \triangle = \bigcirc, \quad \bigcirc = \triangle \times \square, \quad \square \overline{)\dfrac{\triangle}{\bigcirc}}, \quad \triangle \overline{)\dfrac{\square}{\bigcirc}}, \quad \dfrac{\square}{\dfrac{\bigcirc}{\triangle}},$$

$$\triangle \dfrac{\bigcirc}{\square}$$

$$\square = \bigcirc \times \dfrac{1}{\triangle}, \quad \triangle = \bigcirc \times \dfrac{1}{\square}, \quad \dfrac{1}{\square} = \dfrac{\triangle}{\bigcirc}, \quad \dfrac{1}{\triangle} = \dfrac{\square}{\bigcirc}, \quad \dfrac{\bigcirc}{\triangle} = \dfrac{\square}{1},$$

$$\dfrac{\bigcirc}{\square} = \dfrac{\triangle}{1}$$

$$\square \times \triangle = \bigcirc, \quad \bigcirc = \triangle \times \square, \quad \square \overline{)\dfrac{\triangle}{\bigcirc}}, \quad \triangle \overline{)\dfrac{\square}{\bigcirc}}, \quad \dfrac{\square}{\dfrac{\bigcirc}{\triangle}},$$

$$\triangle \dfrac{\bigcirc}{\square}$$

$$\square = \bigcirc \times \dfrac{1}{\triangle}, \quad \triangle = \bigcirc \times \dfrac{1}{\square}, \quad \dfrac{1}{\square} = \dfrac{\triangle}{\bigcirc}, \quad \dfrac{1}{\triangle} = \dfrac{\square}{\bigcirc}, \quad \dfrac{\bigcirc}{\triangle} = \dfrac{\square}{1},$$

$$\dfrac{\bigcirc}{\square} = \dfrac{\triangle}{1}$$

Learning multiplication with multi-concept and multi-format

c	9	8	7
b	2	8	6
a	3	4	5
	1	2	3

The original square is at b2 = □.

$$\square \times \triangle = \bigcirc, \quad \bigcirc = \triangle \times \square, \quad \square \overline{)\dfrac{\triangle}{\bigcirc}}, \quad \triangle \overline{)\dfrac{\square}{\bigcirc}}, \quad \square \dfrac{\bigcirc}{\triangle},$$

$$\triangle \dfrac{\bigcirc}{\square}$$

$$\square = \bigcirc \times \dfrac{1}{\triangle}, \quad \triangle = \bigcirc \times \dfrac{1}{\square}, \quad \dfrac{1}{\square} = \dfrac{\triangle}{\bigcirc}, \quad \dfrac{1}{\triangle} = \dfrac{\square}{\bigcirc}, \quad \dfrac{\bigcirc}{\triangle} = \dfrac{\square}{1},$$

$$\dfrac{\bigcirc}{\square} = \dfrac{\triangle}{1}$$

$$\square \times \triangle = \bigcirc, \quad \bigcirc = \triangle \times \square, \quad \square \overline{)\dfrac{\triangle}{\bigcirc}}, \quad \triangle \overline{)\dfrac{\square}{\bigcirc}}, \quad \square \dfrac{\bigcirc}{\triangle},$$

$$\triangle \dfrac{\bigcirc}{\square}$$

$$\square = \bigcirc \times \dfrac{1}{\triangle}, \quad \triangle = \bigcirc \times \dfrac{1}{\square}, \quad \dfrac{1}{\square} = \dfrac{\triangle}{\bigcirc}, \quad \dfrac{1}{\triangle} = \dfrac{\square}{\bigcirc}, \quad \dfrac{\bigcirc}{\triangle} = \dfrac{\square}{1},$$

$$\dfrac{\bigcirc}{\square} = \dfrac{\triangle}{1}$$

2007 - 2017 © Frank Ho, Amanda Ho, All rights reserved.   www.homathchess.com

Learning multiplication with multi-concept and multi-format

c	9	8	7
b	2	8	6
a	3	4	5
	1	2	3

The original square is at b2 = □.

Student's Name _____ Date _____

2007 - 2017 © Frank Ho, Amanda Ho, All rights reserved.　www.homathchess.com

Learning multiplication with multi-concept and multi-format

c	9	8	7
b	2	9	6
a	3	4	5
	1	2	3

The original square is at b2 = □.

□ × ↔(↕) = _9 × 2_ = 18

□ × ↔(↕) = _9 × 6_ = 54

□ × ↔(↕) = _9 × 4_ = 36

□ × ↔(↕) = _9 × 2_ = 18

□ × ✕ = _9 × 7_ = 63

□ × ✕ = _9 × 5_ = 45

□ × ✕ = _9 × 3_ = 27

□ × ✕ = _9 × 9_ = 81

↔(↕)

$$\square \times \triangle = \bigcirc, \quad \bigcirc = \triangle \times \square, \quad \square \overline{)\dfrac{}{\bigcirc}}, \quad \triangle \overline{)\dfrac{}{\bigcirc}}, \quad \dfrac{\square}{\triangle} \square \overline{)\bigcirc}, \quad \triangle \overline{)\dfrac{\bigcirc}{\square}}$$

$$\square = \bigcirc \times \dfrac{1}{\triangle}, \quad \triangle = \bigcirc \times \dfrac{1}{\square}, \quad \dfrac{1}{\square} = \dfrac{\triangle}{\bigcirc}, \quad \dfrac{1}{\square} = \dfrac{\square}{\triangle}, \quad \dfrac{\bigcirc}{\triangle} = \dfrac{\square}{1}, \quad \dfrac{\bigcirc}{\square} = \dfrac{\triangle}{1}$$

Mom! I Learn Multiplication Using Math-Chess-Puzzles Connection!

Student's Name _____ Date _____

2007 - 2017 © Frank Ho, Amanda Ho, All rights reserved.　　www.homathchess.com

Learning multiplication with multi-concept and multi-format

c	9	8	7
b	2	9	6
a	3	4	5
	1	2	3

The original square is at b2 = □.

Ho Math Chess　何数棋谜　妈！我会棋谜式乘法啦！
Mom! I Learn Multiplication Using Math-Chess-Puzzles Connection!

Student's Name _____ Date _____

2007 - 2017 © Frank Ho, Amanda Ho, All rights reserved.　　www.homathchess.com

Learning multiplication with multi-concept and multi-format

c	9	8	7
b	2	9	6
a	3	4	5
	1	2	3

The original squ
are is at b2 = □.

Ho Math Chess　何数棋谜　妈！我会棋谜式乘法啦！
Mom! I Learn Multiplication Using Math-Chess-Puzzles Connection!

Student's Name _____ Date _____

2007 - 2017 © Frank Ho, Amanda Ho, All rights reserved.　www.homathchess.com

Learning multiplication with multi-concept and multi-format

c	9	8	7
b	2	9	6
a	3	4	5
	1	2	3

The original square is at b2 = □.

Ho Math Chess   何数棋谜   妈！我会棋谜式乘法啦！
Mom! I Learn Multiplication Using Math-Chess-Puzzles Connection!

Student's Name _____ Date _____

2007 - 2017 © Frank Ho, Amanda Ho, All rights reserved.   www.homathchess.com

Learning multiplication with multi-concept and multi-format

c	9	8	7
b	2	9	6
a	3	4	5
	1	2	3

The original square is at b2 = □.

$$\square \times \triangle = \bigcirc, \quad \bigcirc = \triangle \times \square, \quad \square\overline{)\bigcirc}, \quad \triangle\overline{)\bigcirc}, \quad \dfrac{\square \, \bigcirc}{\triangle}, \quad \dfrac{\triangle \, \bigcirc}{\square}$$

$$\square = \bigcirc \times \dfrac{1}{\triangle}, \quad \triangle = \bigcirc \times \dfrac{1}{\square}, \quad \dfrac{1}{\square} = \dfrac{\triangle}{\bigcirc}, \quad \dfrac{1}{\triangle} = \dfrac{\square}{\bigcirc}, \quad \dfrac{\bigcirc}{\triangle} = \dfrac{\square}{1},$$

$$\dfrac{\bigcirc}{\square} = \dfrac{\triangle}{1}$$

2007 - 2017 © Frank Ho, Amanda Ho, All rights reserved.   www.homathchess.com

Intelligent worksheets of students directed Multiplication, addition, and subtraction

3	1	2	3
2	4	5	6
1	7	8	9
	a	b	c

You are at b2 = ☐ .

3	10	30	20
2	15	40	25
1	5	45	35
	d	e	f

☐ × ✛ + or — (circle one) _____ = (shaded square on the right)

__5__ × __6__ + or ⊖(circle one) __5__ = **25**

3			
2			X
1			
	d	e	f

☐ × ✛ + or — (circle one) _____ = (shaded square on the right)

__5__ × __8__ ⊕ or — (circle one) __5__ = **45**

3			
2			
1		X	
	d	e	f

269

Student's Name _____ Date _____

2007 - 2017 © Frank Ho, Amanda Ho, All rights reserved.   www.homathchess.com

Intelligent worksheets of students directed multiplication, addition, and subtraction

3	1	2	3
2	4	5	6
1	7	8	9
	a	b	c

You are at b2 = ☐ .

3	10	30	20
2	15	40	25
1	5	45	35
	d	e	f

☐ × ⬌ + or — (circle one) _____ = (shaded square on the right)

_____ × ____ + or — (circle one) _____ = _____

☐ × ⬌ + or — (circle one) _____ = (shaded square on the right)

_____ × ____ + or — (circle one) _____ = _____

$5 × 4 - 5 = 15$

$5 × 2 + 20 = 30$

Ho Math Chess　何数棋谜　妈！我会棋谜式乘法啦！
Mom! I Learn Multiplication Using Math-Chess-Puzzles Connection!

Student's Name _____ Date _____

2007 - 2017 © Frank Ho, Amanda Ho, All rights reserved.　　www.homathchess.com

Intelligent worksheets of students directed multiplication, addition, and subtraction

3	1	2	3
2	4	5	6
1	7	8	9
	a	b	c

You are at b2 = ☐ .

3	10	30	20
2	15	40	25
1	5	45	35
	d	e	f

☐ × ✕ ＋ or － (circle one) _____ = (shaded square on the right)

_____ × ____ ＋ or － (circle one) _____ = _____

3	✕		
2			
1			
	d	e	f

☐ × ✕ ＋ or － (circle one) _____ = (shaded square on the right)

_____ × ____ ＋ or － (circle one) _____ = _____

3			✕
2			
1			
	d	e	f

5×1+5=10
5×3+5=20

Ho Math Chess   何数棋谜   妈！我会棋谜式乘法啦！
Mom! I Learn Multiplication Using Math-Chess-Puzzles Connection!

Student's Name _____ Date _____

2007 - 2017 © Frank Ho, Amanda Ho, All rights reserved.   www.homathchess.com

Intelligent worksheets of students directed multiplication, addition, and subtraction

3	1	2	3
2	4	5	6
1	7	8	9
	a	b	c

You are at b2 = ☐ .

3	10	30	20
2	15	40	25
1	5	45	35
	d	e	f

☐ × ✕ + or − (circle one) _____ = (shaded square on the right)

_____ × ____ + or − (circle one) _____ = _____

☐ × ✕ + or − (circle one) _____ = (shaded square on the right)

_____ × ____ + or − (circle one) _____ = _____

5 ✕ 9-10=35

5 ✕ 7-3=5

Ho Math Chess 何数棋谜 妈！我会棋谜式乘法啦！
Mom! I Learn Multiplication Using Math-Chess-Puzzles Connection!

Student's Name _____ Date _____

2007 - 2017 © Frank Ho, Amanda Ho, All rights reserved.   www.homathchess.com

Intelligent worksheets of students directed multiplication, addition, and subtraction

3	1	2	3
2	4	6	6
1	7	8	9
	a	b	c

You are at b2 = ☐.

3	54	12	20
2	48	30	18
1	36	24	42
	d	e	f

☐ × ⬌ + or — (circle one) _____ = (shaded square on the right)

_____ × _____ + or — (circle one) _____ = _____

3			
2			x
1			
	d	e	f

☐ × ⬌ + or — (circle one) _____ = (shaded square on the right)

_____ × _____ + or — (circle one) _____ = _____

3			
2			
1		x	
	d	e	f

$6 \times 6 - 18 = 18$
$6 \times 8 - 24 = 24$

2007 - 2017 © Frank Ho, Amanda Ho, All rights reserved.   www.homathchess.com

Intelligent worksheets of students directed multiplication, addition, and subtraction

3	1	2	3
2	4	6	6
1	7	8	9
	a	b	c

You are at b2 = ☐ .

3	54	12	20
2	48	30	18
1	36	24	42
	d	e	f

☐ × ⬌ + or − (circle one) _____ = (shaded square on the right)

3			
2	X		
1			
	d	e	f

_____ × ____ + or − (circle one) _____ = _____

☐ × ⬌ + or − (circle one) _____ = (shaded square on the right)

3		X	
2			
1			
	d	e	f

_____ × ____ + or − (circle one) _____ = _____

$6 \times 4 +_{24=48}$

$6 \times 2 +- 0 = 12$

Ho Math Chess 何数棋谜 妈！我会棋谜式乘法啦！
Mom! I Learn Multiplication Using Math-Chess-Puzzles Connection!

Student's Name _____ Date _____

2007 - 2017 © Frank Ho, Amanda Ho, All rights reserved.   www.homathchess.com

Intelligent worksheets of students directed multiplication, addition, and subtraction

3	1	2	3
2	4	6	6
1	7	8	9
	a	b	c

You are at b2 = ☐.

3	54	12	20
2	48	30	18
1	36	24	42
	d	e	f

☐ × ╳ + or − (circle one) _____ = (shaded square on the right)

_____ × _____ + or − (circle one) _____ = _____

3	x		
2			
1			
	d	e	f

.
☐ × ╳ + or − (circle one) _____ = (shaded square on the right)

_____ × _____ + or − (circle one) _____ = _____

3			x
2			
1			
	d	e	f

6×1+48=54
6×3+2=20

Ho Math Chess 何数棋谜 妈！我会棋谜式乘法啦！
Mom! I Learn Multiplication Using Math-Chess-Puzzles Connection!

Student's Name _____ Date _____

2007 - 2017 © Frank Ho, Amanda Ho, All rights reserved.   www.homathchess.com

Intelligent worksheets of students directed multiplication, addition, and subtraction

3	1	2	3
2	4	6	6
1	7	8	9
	a	b	c

You are at b2 = ☐ .

3	54	12	20
2	48	30	18
1	36	24	42
	d	e	f

☐ × ╳ + or − (circle one) _____ = (shaded square on the right)

_____ × _____ + or − (circle one) _____ = _____

3			
2			
1			▨
	d	e	f

☐ × ╳ + or − (circle one) _____ = (shaded square on the right)

_____ × _____ + or − (circle one) _____ = _____

3			
2			
1	▨		
	d	e	f

6×9-12=42
6×7-6=36

Intelligent worksheets of students directed multiplication, addition, and subtraction

3	1	2	3
2	4	7	6
1	7	8	9
	a	b	c

You are at b2 = ☐.

3	31	30	63
2	15	28	56
1	49	42	35
	d	e	f

☐ × ⬌ + or − (circle one) _____ = (shaded square on the right)

_____ × ____ + or − (circle one) _____ = _____

3			
2			X
1			
	d	e	f

☐ × ⬍ + or − (circle one) _____ = (shaded square on the right)

_____ × ____ + or − (circle one) _____ = _____

3			
2			
1		X	
	d	e	f

7×6 + 14=56
7×8-14=42

277

Ho Math Chess　何数棋谜　妈！我会棋谜式乘法啦！
Mom! I Learn Multiplication Using Math-Chess-Puzzles Connection!

Student's Name _____ Date _____

2007 - 2017 © Frank Ho, Amanda Ho, All rights reserved.　www.homathchess.com

Intelligent worksheets of students directed multiplication, addition, and subtraction

3	1	2	3
2	4	7	6
1	7	8	9
	a	b	c

You are at b2 = ☐ .

3	31	30	63
2	15	28	56
1	49	42	35
·d	e	f

☐ × ✛ + or — (circle one) _____ = (shaded square on the right)

_____ × ____ + or — (circle one) _____ = _____

3			
2	x		
1			
	d	e	f

☐ × ✛ + or — (circle one) _____ = (shaded square on the right)

_____ × ____ + or — (circle one) _____ = _____

3		x	
2			
1			
	d	e	f

$7 \times 4 - 13 = 15$

$7 \times 2 +_{16} = 30$

Student's Name _____ Date _____

2007 - 2017 © Frank Ho, Amanda Ho, All rights reserved.   www.homathchess.com

Intelligent worksheets of students directed multiplication, addition, and subtraction

3	1	2	3
2	4	7	6
1	7	8	9
	a	b	c

You are at b2 = ☐.

3	31	30	63
2	15	28	56
1	49	42	35
	d	e	f

☐ × ╳ + or − (circle one) _____ = (shaded square on the right)

_____ × ____ + or − (circle one) _____ = _____

☐ × ╳ + or − (circle one) _____ = (shaded square on the right)

_____ × ____ + or − (circle one) _____ = _____

7 ╳ 1+24=31

7×3+42=63

279

Ho Math Chess　何数棋谜　妈！我会棋谜式乘法啦！

Mom! I Learn Multiplication Using Math-Chess-Puzzles Connection!

Student's Name _____ Date _____

2007 - 2017 © Frank Ho, Amanda Ho, All rights reserved.　　www.homathchess.com

Intelligent worksheets of students directed multiplication, addition, and subtraction

3	1	2	3
2	4	7	6
1	7	8	9
	a	b	c

You are at b2 = ☐ .

3	31	30	63
2	15	28	56
1	49	42	35
	d	e	f

☐ × ✕ + or − (circle one) _____ = (shaded square on the right)

_____ × _____ + or − (circle one) _____ = _____

3			
2			
1			X
	d	e	f

☐ × ✕ + or − (circle one) _____ = (shaded square on the right)

_____ × _____ + or − (circle one) _____ = _____

3			
2			
1	X		
	d	e	f

$7×9-38=35$

$7×7+- 0 =49$

Ho Math Chess   何数棋谜   妈！我会棋谜式乘法啦！

Mom! I Learn Multiplication Using Math-Chess-Puzzles Connection!

Student's Name _____ Date _____

2007 - 2017 © Frank Ho, Amanda Ho, All rights reserved.   www.homathchess.com

Intelligent worksheets of students directed multiplication, addition, and subtraction

3	1	2	3
2	4	8	6
1	7	8	9
	a	b	c

You are at b2 = ☐ .

3	48	56	24
2	64	40	72
1	24	32	16
	d	e	f

☐ × ⬍ + or — (circle one) _____ = (shaded square on the right)

_____ × ____ + or — (circle one) _____ = _____

☐ × ⬍ + or — (circle one) _____ = (shaded square on the right)

_____ × ____ + or — (circle one) _____ = _____

8×6+24=72
8×8-32=32

Ho Math Chess   何数棋谜   妈!我会棋谜式乘法啦!
Mom! I Learn Multiplication Using Math-Chess-Puzzles Connection!

Student's Name _____ Date _____

2007 - 2017 © Frank Ho, Amanda Ho, All rights reserved.   www.homathchess.com

Intelligent worksheets of students directed multiplication, addition, and subtraction

3	1	2	3
2	4	8	6
1	7	8	9
	a	b	c

You are at b2 = ☐.

3	48	56	24
2	64	40	72
1	24	32	16
	d	e	f

☐ × ⬍ + or — (circle one) _____ = (shaded square on the right)

_____ × ____ + or — (circle one) _____ = _____

☐ × ⬍ + or — (circle one) _____ = (shaded square on the right)

_____ × ____ + or — (circle one) _____ = _____

8×4+32=64

$8 \times 2 + 40 = 56$

2007 - 2017 © Frank Ho, Amanda Ho, All rights reserved.   www.homathchess.com

Intelligent worksheets of students directed multiplication, addition, and subtraction

3	1	2	3
2	4	8	6
1	7	8	9
	a	b	c

You are at b2 = ☐.

3	48	56	24
2	64	40	72
1	24	32	16
	d	e	f

☐ × ✕ ✛ or ➖ (circle one) _____ = (shaded square on the right)

_____ × _____ ✛ or ➖ (circle one) _____ = _____

☐ × ✕ ✛ or ➖ (circle one) _____ = (shaded square on the right)

_____ × _____ ✛ or ➖ (circle one) _____ = _____

$8 \times 1 + 40 = 48$

$8 \times 3 + - 0 = 24$

2007 - 2017 © Frank Ho, Amanda Ho, All rights reserved.　　www.homathchess.com

Intelligent worksheets of students directed multiplication, addition, and subtraction

3	1	2	3
2	4	8	6
1	7	8	9
	a	b	c

You are at b2 = ☐.

3	48	56	24
2	64	40	72
1	24	32	16
	d	e	f

☐ × ✕ + or − (circle one) _____ = (shaded square on the right)

_____ × _____ + or − (circle one) _____ = _____

☐ × ✕ + or − (circle one) _____ = (shaded square on the right)

_____ × _____ + or − (circle one) _____ = _____

$8 \times 9 - 56 = 16$

$8 \times 7 - 32 = 24$

Ho Math Chess 何数棋谜 妈！我会棋谜式乘法啦！

Mom! I Learn Multiplication Using Math-Chess-Puzzles Connection!

Student's Name _____ Date _____

2007 - 2017 © Frank Ho, Amanda Ho, All rights reserved.   www.homathchess.com

Intelligent worksheets of students directed multiplication, addition, and subtraction

3	1	2	3
2	4	9	6
1	7	8	9
	a	b	c

You are at b2 = ☐.

3	36	81	27
2	63	40	54
1	18	45	36
	d	e	f

☐ × ⬍ + or — (circle one) _____ = (shaded square on the right)

_____ × ____ + or — (circle one) _____ = _____

3			
2			x
1			
	d	e	f

☐ × ⬍ + or — (circle one) _____ = (shaded square on the right)

_____ × ____ + or — (circle one) _____ = _____

3			
2			
1		x	
	d	e	f

9 × 6 + - 0 = 54

7 × 8 – 27 = 45

Ho Math Chess 何数棋谜 妈！我会棋谜式乘法啦！
Mom! I Learn Multiplication Using Math-Chess-Puzzles Connection!

Student's Name _____ Date _____

2007 - 2017 © Frank Ho, Amanda Ho, All rights reserved.   www.homathchess.com

Intelligent worksheets of students directed multiplication, addition, and subtraction

3	1	2	3
2	4	9	6
1	7	8	9
	a	b	c

You are at b2 = ☐.

3	36	81	27
2	63	40	54
1	18	45	36
	d	e	f

☐ × ⊕ + or − (circle one) _____ = (shaded square on the right)

_____ × _____ + or − (circle one) _____ = _____

☐ × ⊕ + or − (circle one) _____ = (shaded square on the right)

_____ × _____ + or − (circle one) _____ = _____

9 × 4 + 27 = 63

9 × 2 + 63 = 81

Mom! I Learn Multiplication Using Math-Chess-Puzzles Connection!

Student's Name _____ Date _____

2007 - 2017 © Frank Ho, Amanda Ho, All rights reserved.   www.homathchess.com

Intelligent worksheets of students directed multiplication, addition, and subtraction

3	1	2	3
2	4	9	6
1	7	8	9
	a	b	c

You are at b2 = ☐.

3	36	81	27
2	63	40	54
1	18	45	36
	d	e	f

☐ × ✕ + or − (circle one) _____ = (shaded square on the right)

_____ × _____ + or − (circle one) _____ = _____

3	X		
2			
1			
	d	e	f

☐ × ✕ + or − (circle one) _____ = (shaded square on the right)

_____ × _____ + or − (circle one) _____ = _____

3			X
2			
1			
	d	e	f

9 ✕ 1 + 25 = 36

9 ✕ 3 + − 0 = 27

Ho Math Chess   何数棋谜   妈！我会棋谜式乘法啦！
Mom! I Learn Multiplication Using Math-Chess-Puzzles Connection!

Student's Name _____ Date _____

2007 - 2017 © Frank Ho, Amanda Ho, All rights reserved.   www.homathchess.com

Intelligent worksheets of students directed multiplication, addition, and subtraction

3	1	2	3
2	4	9	6
1	7	8	9
	a	b	c

You are at b2 = ☐ .

3	36	81	27
2	63	40	54
1	18	45	36
	d	e	f

☐ × ╳ + or − (circle one) _____ = (shaded square on the right)

_____ × _____ + or − (circle one) _____ = _____

3			
2			
1			x
	d	e	f

☐ × ╳ + or − (circle one) _____ = (shaded square on the right)

_____ × _____ + or − (circle one) _____ = _____

3			
2			
1	x		
	d	e	f

9 ╳ 9 − 45 = 36

9 ╳ 7 − 45 = 18

Student's Name _____ Date _____

2007 - 2017 © Frank Ho, Amanda Ho, All rights reserved.   www.homathchess.com

Intelligent worksheets of students directed multiplication, addition, and subtraction

3	1	2	3
2	4	4	6
1	7	8	9
	a	b	c

3	36	27	12
2	24	28	8
1	20	32	16
	d	e	f

You are at b2 = ☐.

☐ × ✛ + or — (circle one) _____ = (shaded square on the right)

_____ × ____ + or — (circle one) _____ = _____

3			
2			▨
1			
	d	e	f

☐ × ✛ + or — (circle one) _____ = (shaded square on the right)

_____ × ____ + or — (circle one) _____ = _____

3			
2			
1		▨	
	d	e	f

$4 \times 6 - 16 = 8$

$4 \times 8 + - _0 = 32$

Ho Math Chess   何数棋谜   妈！我会棋谜式乘法啦！

Mom! I Learn Multiplication Using Math-Chess-Puzzles Connection!

Student's Name _____ Date _____

2007 - 2017 © Frank Ho, Amanda Ho, All rights reserved.   www.homathchess.com

Intelligent worksheets of students directed multiplication, addition, and subtraction

3	1	2	3
2	4	4	6
1	7	8	9
	a	b	c

You are at b2 = ☐ .

3	36	27	12
2	24	28	8
1	20	32	16
	d	e	f

☐ × ✛ + or — (circle one) _____ = (shaded square on the right)

_____ × ____ + or — (circle one) _____ = _____

☐ × ✛ + or — (circle one) _____ = (shaded square on the right)

_____ × ____ + or — (circle one) _____ = _____

4 × 4+8=24

4 × 2+19=27

Mom! I Learn Multiplication Using Math-Chess-Puzzles Connection!

Student's Name _____ Date _____

2007 - 2017 © Frank Ho, Amanda Ho, All rights reserved.　　www.homathchess.com

Intelligent worksheets of students directed multiplication, addition, and subtraction

3	1	2	3
2	4	4	6
1	7	8	9
	a	b	c

You are at b2 = ☐.

3	36	27	12
2	24	28	8
1	20	32	16
	d	e	f

☐ × ✕ **+** or **–** (circle one) _____ = (shaded square on the right)

3	X		
2			
1			
	d	e	f

_____ × ____ **+** or **–** (circle one) _____ = _____

☐ × ✕ **+** or **–** (circle one) _____ = (shaded square on the right)

3			X
2			
1			
	d	e	f

_____ × ____ **+** or **–** (circle one) _____ = _____

$4 ✕ 1 + 32 = 36$

$4 ✕ 3 + - 0 = 12$

Ho Math Chess 何数棋谜 妈！我会棋谜式乘法啦！
Mom! I Learn Multiplication Using Math-Chess-Puzzles Connection!

Student's Name _____ Date _____

2007 - 2017 © Frank Ho, Amanda Ho, All rights reserved.   www.homathchess.com

Intelligent worksheets of students directed multiplication, addition, and subtraction

3	1	2	3
2	4	4	6
1	7	8	9
	a	b	c

You are at b2 = ☐ .

3	36	27	12
2	24	28	8
1	20	32	16
	d	e	f

☐ × ✕ + or − (circle one) _____ = (shaded square on the right)

_____ × _____ + or − (circle one) _____ = _____

3			
2			
1			X
	d	e	f

☐ × ✕ + or − (circle one) _____ = (shaded square on the right)

_____ × _____ + or − (circle one) _____ = _____

3			
2			
1	X		
	d	e	f

4 ✕ 9-20=16

4 ✕ 7-8=20

Ho Math Chess   何数棋谜   妈！我会棋谜式乘法啦！
Mom! I Learn Multiplication Using Math-Chess-Puzzles Connection!

Student's Name _____ Date _____

2007 - 2017 © Frank Ho, Amanda Ho, All rights reserved.   www.homathchess.com

Intelligent worksheets of students directed multiplication, addition, and subtraction

3	1	2	3
2	4	3	6
1	7	8	9
	a	b	c

You are at b2 = ☐.

3	6	27	12
2	15	24	21
1	12	9	18
	d	e	f

☐ × ✛ + or − (circle one) _____ = (shaded square on the right)

_____ × _____ + or − (circle one) _____ = _____

3			
2			X
1			
	d	e	f

☐ × ✛ + or − (circle one) _____ = (shaded square on the right)

_____ × _____ + or − (circle one) _____ = _____

3			
2			
1		X	
	d	e	f

3 × 6 + 3 = 21

3 × 8 - 15 = 9

Ho Math Chess 何数棋谜 妈!我会棋谜式乘法啦!

Mom! I Learn Multiplication Using Math-Chess-Puzzles Connection!

Student's Name _____ Date _____

2007 - 2017 © Frank Ho, Amanda Ho, All rights reserved.   www.homathchess.com

Intelligent worksheets of students directed multiplication, addition, and subtraction

3	1	2	3
2	4	3	6
1	7	8	9
	a	b	c

You are at b2 = ▢ .

3	6	27	12
2	15	24	21
1	12	9	18
	d	e	f

▢ × ✥ + or — (circle one) _____ = (shaded square on the right)

_____ × ____ + or — (circle one) _____ = _____

3			
2	X		
1			
	d	e	f

▢ × ✥ + or — (circle one) _____ = (shaded square on the right)

_____ × ____ + or — (circle one) _____ = _____

3		X	
2			
1			
	d	e	f

3 × 4 + 3 = 15

3 × 2 + 21 = 27

Mom! I Learn Multiplication Using Math-Chess-Puzzles Connection!

Student's Name _____ Date _____

2007 - 2017 © Frank Ho, Amanda Ho, All rights reserved.　　www.homathchess.com

Intelligent worksheets of students directed multiplication, addition, and subtraction

3	1	2	3
2	4	3	6
1	7	8	9
	a	b	c

You are at b2 = ☐ .

3	6	27	12
2	15	24	21
1	12	9	18
	d	e	f

☐ × ✕ + or − (circle one) _____ = (shaded square on the right)

_____ × _____ + or − (circle one) _____ = _____

3	✗		
2			
1			
	d	e	f

☐ × ✕ + or − (circle one) _____ = (shaded square on the right)

_____ × _____ + or − (circle one) _____ = _____

3			✗
2			
1			
	d	e	f

$3 ✕ 1 + 3 = 6$

$3 ✕ 3 + 3 = 12$

Ho Math Chess　何数棋谜　妈！我会棋谜式乘法啦！
Mom! I Learn Multiplication Using Math-Chess-Puzzles Connection!

Student's Name _____ Date _____

2007 - 2017 © Frank Ho, Amanda Ho, All rights reserved.　　www.homathchess.com

Intelligent worksheets of students directed multiplication, addition, and subtraction

3	1	2	3
2	4	3	6
1	7	8	9
	a	b	c

You are at b2 = ☐ .

3	6	27	12
2	15	24	21
1	12	9	18
	d	e	f

☐ × ✕ + or − (circle one) _____ = (shaded square on the right)

_____ × ____ + or − (circle one) _____ = _____

☐ × ✕ + or − (circle one) _____ = (shaded square on the right)

_____ × ____ + or − (circle one) _____ = _____

$3 × 9 - 9 = 18$

$_3 × 7 - 9 = 12$

Ho Math Chess   何数棋谜   妈！我会棋谜式乘法啦！
Mom! I Learn Multiplication Using Math-Chess-Puzzles Connection!

Student's Name _____ Date _____

2007 - 2017 © Frank Ho, Amanda Ho, All rights reserved.   www.homathchess.com

Multiplication table

×	200	201	202	203	204	205	206	207	208	209
♙	200	201	202	203	204	205	206	207	208	209
2	400	402	404	406	408	410	412	414	416	418
♗	600	603	606	609	612	615	618	621	624	627
4	800	804	808	812	816	820	824	828	832	836
♖	1000	1005	1010	1015	1020	1025	1030	1035	1040	1045
6	1200	1206	1212	1218	1224	1230	1236	1242	1248	1254
7	1400	1407	1414	1421	1428	1435	1442	1449	1456	1463
8	1600	1608	1616	1624	1632	1640	1648	1656	1664	1672
♕	1800	1809	1818	1827	1836	1845	1854	1863	1872	1881

2007 - 2017 © Frank Ho, Amanda Ho, All rights reserved.　www.homathchess.com

Multiplication table

×	300	301	302	303	304	305	306	307	308	309
♙	300	301	302	303	304	305	306	307	308	309
2	600	602	604	606	608	610	612	614	616	618
♗	900	903	906	909	912	915	918	921	924	927
4	1200	1204	1208	1212	1216	1220	1224	1228	1232	1236
♖	1500	1505	1510	1515	1520	1525	1530	1535	1540	1545
6	1800	1806	1812	1818	1824	1830	1836	1842	1848	1854
7	2100	2107	2114	2121	2128	2135	2142	2149	2156	2163
8	2400	2408	2416	2424	2432	2440	2448	2456	2464	2472
♛	2700	2709	2718	2727	2736	2745	2754	2763	2772	2781

2007 - 2017 © Frank Ho, Amanda Ho, All rights reserved.　　www.homathchess.com

Multiplication table

×	400	401	402	403	404	405	406	407	408	409
♟	400	401	402	403	404	405	406	407	408	409
2	800	802	804	806	808	810	812	814	816	818
♝	1200	1203	1206	1209	1212	1215	1218	1221	1224	1227
4	1600	1604	1608	1612	1616	1620	1624	1628	1632	1636
♜	2000	2005	2010	2015	2020	2025	2030	2035	2040	2045
6	2400	2406	2412	2418	2424	2430	2436	2442	2448	2454
7	2800	2807	2814	2821	2828	2835	2842	2849	2856	2863
8	3200	3208	3216	3224	3232	3240	3248	3256	3264	3272
♛	3600	3609	3618	3627	3636	3645	3654	3663	3672	3681

Ho Math Chess　何数棋谜　妈！我会棋谜式乘法啦！
Mom! I Learn Multiplication Using Math-Chess-Puzzles Connection!

Student's Name _____ Date _____

2007 - 2017 © Frank Ho, Amanda Ho, All rights reserved.　www.homathchess.com

Multiplication table

×	500	501	502	503	504	505	506	507	508	509
♙	500	501	502	503	504	505	506	507	508	509
2	1000	1002	1004	1006	1008	1010	1012	1014	1016	1018
♗	1500	1503	1506	1509	1512	1515	1518	1521	1524	1527
4	2000	2004	2008	2012	2016	2020	2024	2028	2032	2036
♖	2500	2505	2510	2515	2520	2525	2530	2535	2540	2545
6	3000	3006	3012	3018	3024	3030	3036	3042	3048	3054
7	3500	3507	3514	3521	3528	3535	3542	3549	3556	3563
8	4000	4008	4016	4024	4032	4040	4048	4056	4064	4072
♕	4500	4509	4518	4527	4536	4545	4554	4563	4572	4581

Mom! I Learn Multiplication Using Math-Chess-Puzzles Connection!

Student's Name _____ Date _____

2007 - 2017 © Frank Ho, Amanda Ho, All rights reserved.   www.homathchess.com

Multiplication table

✕	600	601	602	603	604	605	606	607	608	609
♟	600	601	602	603	604	605	606	607	608	609
2	1200	1202	1204	1206	1208	1210	1212	1214	1216	1218
♗	1800	1803	1806	1809	1812	1815	1818	1821	1824	1827
4	2400	2404	2408	2412	2416	2420	2424	2428	2432	2436
♖	3000	3005	3010	3015	3020	3025	3030	3035	3040	3045
6	3600	3606	3612	3618	3624	3630	3636	3642	3648	3654
7	4200	4207	4214	4221	4228	4235	4242	4249	4256	4263
8	4800	4808	4816	4824	4832	4840	4848	4856	4864	4872
♕	5400	5409	5418	5427	5436	5445	5454	5463	5472	5481

Multiplication table

×	700	701	702	703	704	705	706	707	708	709
♙	700	701	702	703	704	705	706	707	708	709
2	1400	1402	1404	1406	1408	1410	1412	1414	1416	1418
♗	2100	2103	2106	2109	2112	2115	2118	2121	2124	2127
4	2800	2804	2808	2812	2816	2820	2824	2828	2832	2836
♖	3500	3505	3510	3515	3520	3525	3530	3535	3540	3545
6	4200	4206	4212	4218	4224	4230	4236	4242	4248	4254
7	4900	4907	4914	4921	4928	4935	4942	4949	4956	4963
8	5600	5608	5616	5624	5632	5640	5648	5656	5664	5672
♕	6300	6309	6318	6327	6336	6345	6354	6363	6372	6381

2007 - 2017 © Frank Ho, Amanda Ho, All rights reserved.   www.homathchess.com

Multiplication table

×	800	801	802	803	804	805	806	807	808	809
♙	800	801	802	803	804	805	806	807	808	809
2	1600	1602	1604	1606	1608	1610	1612	1614	1616	1618
♗	2400	2403	2406	2409	2412	2415	2418	2421	2424	2427
4	3200	3204	3208	3212	3216	3220	3224	3228	322	3236
♖	4000	4005	4010	4015	4020	4025	4030	4035	4040	4045
6	4800	4806	4812	4818	4824	4830	4836	4842	4848	4854
7	5600	5607	5614	5621	5628	5635	5642	5649	5656	5663
8	8400	8408	8416	8424	8432	8440	8449	8464	8464	8472
♕	7200	7209	7218	7227	7236	7245	7254	7263	7272	7281

Ho Math Chess　何数棋谜　妈！我会棋谜式乘法啦！
Mom! I Learn Multiplication Using Math-Chess-Puzzles Connection!

Student's Name _____ Date _____
2007 - 2017 © Frank Ho, Amanda Ho, All rights reserved.　　www.homathchess.com

Multiplication table

×	900	901	902	903	904	905	906	907	908	909
♟	900	901	902	903	904	905	906	907	908	909
2	1800	1802	1804	1806	1808	1810	1812	1814	1816	1818
♝	2700	2703	2706	2709	2712	2715	2718	2721	2724	2727
4	3600	3604	3608	3612	3616	3620	3624	3628	3632	3636
♜	4500	4505	4510	4515	4520	4525	4530	4535	4540	4545
6	5400	5406	5412	5418	5424	5430	5436	5442	5448	5454
7	6300	6307	6314	6321	6328	6335	6342	6349	6356	6363
8	7200	7208	7216	7224	7232	7240	7248	7256	7264	7272
♛	8100	819	8118	8127	8136	8145	8154	8163	8172	8181

Ho Math Chess 何数棋谜 妈！我会棋谜式乘法啦！
Mom! I Learn Multiplication Using Math-Chess-Puzzles Connection!

Student's Name _____ Date _____

2007 - 2017 © Frank Ho, Amanda Ho, All rights reserved.   www.homathchess.com

Multiplication table

✕	11	12	13	14	15	16	17	18	19
♙	11	12	13	14	15	16	17	18	19
2	22	24	26	28	30	32	34	36	38
♗	33	36	39	42	45	48	51	54	57
4	44	48	52	56	60	64	68	72	76
♖	55	60	65	70	75	80	85	90	95
6	66	72	78	84	90	96	102	108	114
7	77	84	91	98	105	112	119	126	133
8	88	96	104	112	120	128	136	144	152
♕	99	108	117	126	135	144	153	162	171

2007 - 2017 © Frank Ho, Amanda Ho, All rights reserved.   www.homathchess.com

Multiplication table

✕	111	121	131	141	151	161	171	181	191
♟	111	121	131	141	151	161	171	181	191
2	222	242	262	282	302	322	342	362	382
♝	333	363	393	423	453	483	513	543	573
4	444	484	524	564	604	644	684	724	764
♜	555	605	655	705	755	805	855	905	955
6	666	726	786	846	906	966	1026	1086	1146
7	777	847	917	987	1057	1127	1197	1267	1337
8	888	968	1048	1128	1208	1288	1368	1448	1528
♛	999	1089	1179	1269	1359	1449	1539	1629	1719

2007 - 2017 © Frank Ho, Amanda Ho, All rights reserved.   www.homathchess.com

Multiplication table

×	11	22	33	44	55	66	77	88	99
♙	11	22	33	44	55	66	77	88	99
2	22	44	66	88	110	132	154	176	198
♗	33	66	99	132	165	198	231	264	297
4	44	88	132	176	220	264	308	352	396
♖	55	110	165	220	275	330	385	440	495
6	66	132	198	264	330	396	462	528	594
7	77	154	231	308	385	462	539	616	693
8	88	176	264	352	440	528	616	704	792
♕	99	198	297	396	495	594	693	792	891

Ho Math Chess　何数棋谜　妈！我会棋谜式乘法啦！
Mom! I Learn Multiplication Using Math-Chess-Puzzles Connection!

Student's Name _____ Date _____

2007 - 2017 © Frank Ho, Amanda Ho, All rights reserved.　　www.homathchess.com

dd ✕ d with carrying

¹ 1 2	¹ 1 3	¹ 1 2	¹ 1 2	¹ 1 2
x 6	x 6	x 7	x 8	x ♜
7 2	7 8	8 4	9 6	6 0

1 3	1 3	1 3	1 3	1 3
x ♜	x 6	x 7	x 8	x ♛
6 5	7 8	9 1	1 0 4	1 1 7

² 1 4	¹ 1 4	² 1 4	³ 1 4	² 1 4
x 6	x 4	x 7	x 8	x 5
8 4	5 6	9 8	1 1 2	7 0

1 5	1 5	1 5	1 5	1 5
x ♜	x 6	x 7	x 8	x 9
7 5	9 0	1 0 5	1 2 0	1 3 5

Student's Name _____ Date _____

2007 - 2017 © Frank Ho, Amanda Ho, All rights reserved.　　www.homathchess.com

dd × d with carrying

1	1	1	1	1
2 2	2 3	2 2	2 2	2 2
x 6	x 6	x 7	x 8	x ♜
1 3 2	1 3 8	1 5 4	1 7 6	1 1 0

2 3	2 3	2 3	2 3	2 3
x 5	x 6	x 7	x 8	x 9
1 1 5	1 3 8	1 6 1	1 8 4	2 0 7

2	1	2	3	2
2 4	2 4	2 4	2 4	2 4
x 6	x 4	x 7	x 8	x ♜
1 4 4	9 6	1 6 8	1 9 2	1 2 0

2 5	2 5	2 5	2 5	2 5
x 5	x 6	x 7	x 8	x ♛
1 2 5	1 5 0	1 7 5	2 0 0	2 2 5

2007 - 2017 © Frank Ho, Amanda Ho, All rights reserved.　　www.homathchess.com

dd × d with carrying

1	1	1	1	1
32	33	32	32	32
× 6	× 6	× 7	× 8	× 5
192	198	224	256	160

33	33	33	33	33
× ▨	× 6	× 7	× 8	× ♛
165	198	231	264	297

2	1	2	3	2
34	34	34	34	34
× 6	× 4	× 7	× 8	× ▨
204	136	238	272	170

35	35	35	35	35
× ▨	× 6	× 7	× 8	× ♛
175	210	245	280	315

Ho Math Chess　何数棋谜　妈！我会棋谜式乘法啦！
Mom! I Learn Multiplication Using Math-Chess-Puzzles Connection!

Student's Name _____ Date _____

2007 - 2017 © Frank Ho, Amanda Ho, All rights reserved.　www.homathchess.com

dd × d with carrying

1	1	1	1	1
42	43	42	42	42
x 6	x 6	x 7	x 8	x ▨
252	258	294	336	210

43	43	43	43	43
x 5	x 6	x 7	x 8	x ♛
215	258	301	344	387

2	1	2	3	2
44	44	44	44	44
x 6	x 4	x 7	x 8	x 5
264	176	308	302	220

45	45	45	45	45
x ▨	x 6	x 7	x 8	x ♛
225	270	315	360	405

2007 - 2017 © Frank Ho, Amanda Ho, All rights reserved.   www.homathchess.com

dd × d with carrying

¹ 52	¹ 53	¹ 52	¹ 52	¹ 52
x 6	x 6	x 7	x 8	x ♖
312	318	364	416	260

53	53	53	53	53
x ♖	x 6	x 7	x 8	x ♕
265	318	371	424	477

² 54	¹ 54	² 54	³ 54	² 54
x 6	x 4	x 7	x 8	x 5
324	216	378	432	270

55	55	55	55	55
x ♖	x 6	x 7	x 8	x ♕
275	330	385	440	495

2007 - 2017 © Frank Ho, Amanda Ho, All rights reserved.   www.homathchess.com

dd × d with carrying

¹ 62	¹ 63	¹ 62	¹ 62	¹ 62
× 6	× 6	× 7	× 8	× 5
372	378	434	496	310

63	63	63	63	63
× ♜	× 6	× 7	× 8	× ♛
315	378	441	504	567

² 64	¹ 64	² 64	³ 64	² 64
× 6	× 4	× 7	× 8	× ♜
384	256	448	512	320

65	65	65	65	65
× ♜	× 6	× 7	× 8	× ♛
325	390	455	520	585

2007 - 2017 © Frank Ho, Amanda Ho, All rights reserved.   www.homathchess.com

dd × d with carrying

| $\begin{array}{r}1\\72\\\times\ 6\\\hline 432\end{array}$ | $\begin{array}{r}1\\73\\\times\ 6\\\hline 438\end{array}$ | $\begin{array}{r}1\\72\\\times\ 7\\\hline 504\end{array}$ | $\begin{array}{r}1\\72\\\times\ 8\\\hline 576\end{array}$ | $\begin{array}{r}1\\72\\\times\ ♖\\\hline 360\end{array}$ |

| $\begin{array}{r}73\\\times\ ♖\\\hline 365\end{array}$ | $\begin{array}{r}73\\\times\ 6\\\hline 438\end{array}$ | $\begin{array}{r}73\\\times\ 7\\\hline 511\end{array}$ | $\begin{array}{r}73\\\times\ 8\\\hline 584\end{array}$ | $\begin{array}{r}73\\\times\ ♕\\\hline 657\end{array}$ |

| $\begin{array}{r}2\\74\\\times\ 6\\\hline 444\end{array}$ | $\begin{array}{r}1\\74\\\times\ 4\\\hline 296\end{array}$ | $\begin{array}{r}2\\74\\\times\ 7\\\hline 518\end{array}$ | $\begin{array}{r}3\\74\\\times\ 8\\\hline 592\end{array}$ | $\begin{array}{r}2\\74\\\times\ 5\\\hline 370\end{array}$ |

| $\begin{array}{r}75\\\times\ ♖\\\hline 375\end{array}$ | $\begin{array}{r}75\\\times\ 6\\\hline 450\end{array}$ | $\begin{array}{r}75\\\times\ 7\\\hline 525\end{array}$ | $\begin{array}{r}75\\\times\ 8\\\hline 600\end{array}$ | $\begin{array}{r}75\\\times\ ♕\\\hline 675\end{array}$ |

2007 - 2017 © Frank Ho, Amanda Ho, All rights reserved.   www.homathchess.com

dd × d with carrying

1	1	1	1	1
82	83	82	82	82
× 6	× 6	× 7	× 8	× ♖
492	498	574	656	410

83	83	83	83	83
× ♖	× 6	× 7	× 8	× ♛
415	498	581	664	747

2	1	2	3	2
84	84	84	84	84
× 6	× 4	× 7	× 8	× ♖
504	336	588	672	420

85	85	85	85	85
× ♖	× 6	× 7	× 8	× ♛
425	510	595	680	765

315

Ho Math Chess   何数棋谜 妈！我会棋谜式乘法啦！
Mom! I Learn Multiplication Using Math-Chess-Puzzles Connection!

Student's Name _____ Date _____

2007 - 2017 © Frank Ho, Amanda Ho, All rights reserved.   www.homathchess.com

dd × d with carrying

$$
\begin{array}{r} 1 \\ 92 \\ \times\ 6 \\ \hline 552 \end{array}
\qquad
\begin{array}{r} 1 \\ 93 \\ \times\ 6 \\ \hline 558 \end{array}
\qquad
\begin{array}{r} 1 \\ 92 \\ \times\ 7 \\ \hline 644 \end{array}
\qquad
\begin{array}{r} 1 \\ 92 \\ \times\ 8 \\ \hline 736 \end{array}
\qquad
\begin{array}{r} 1 \\ 92 \\ \times\ ♜ \\ \hline 460 \end{array}
$$

$$
\begin{array}{r} 93 \\ \times\ ♜ \\ \hline 465 \end{array}
\qquad
\begin{array}{r} 93 \\ \times\ 6 \\ \hline 558 \end{array}
\qquad
\begin{array}{r} 93 \\ \times\ 7 \\ \hline 651 \end{array}
\qquad
\begin{array}{r} 93 \\ \times\ 8 \\ \hline 744 \end{array}
\qquad
\begin{array}{r} 93 \\ \times\ ♛ \\ \hline 837 \end{array}
$$

$$
\begin{array}{r} 2 \\ 94 \\ \times\ 6 \\ \hline 564 \end{array}
\qquad
\begin{array}{r} 1 \\ 94 \\ \times\ 4 \\ \hline 376 \end{array}
\qquad
\begin{array}{r} 2 \\ 94 \\ \times\ 7 \\ \hline 658 \end{array}
\qquad
\begin{array}{r} 3 \\ 94 \\ \times\ 8 \\ \hline 752 \end{array}
\qquad
\begin{array}{r} 2 \\ 94 \\ \times\ ♜ \\ \hline 470 \end{array}
$$

$$
\begin{array}{r} 95 \\ \times\ ♜ \\ \hline 475 \end{array}
\qquad
\begin{array}{r} 95 \\ \times\ 6 \\ \hline 570 \end{array}
\qquad
\begin{array}{r} 95 \\ \times\ 7 \\ \hline 665 \end{array}
\qquad
\begin{array}{r} 95 \\ \times\ 8 \\ \hline 760 \end{array}
\qquad
\begin{array}{r} 95 \\ \times\ ♛ \\ \hline 855 \end{array}
$$

Ho Math Chess　何数棋谜　妈！我会棋谜式乘法啦！
Mom! I Learn Multiplication Using Math-Chess-Puzzles Connection!

Student's Name _____ Date _____

2007 - 2017 © Frank Ho, Amanda Ho, All rights reserved.　　www.homathchess.com

Multiplication table

✕	11	11	11	11	11	11	11	11
♙	11	11	11	11	11	11	11	11
2	22	22	22	22	22	22	22	22
♗	33	33	33	33	33	33	33	33
4	44	44	44	44	44	44	44	44
♖	55	55	55	55	55	55	55	55
6	66	66	66	66	66	66	66	66
7	77	77	77	77	77	77	77	77
8	88	88	88	88	88	88	88	88
♕	99	99	99	99	99	99	99	99

2007 - 2017 © Frank Ho, Amanda Ho, All rights reserved.　　www.homathchess.com

Multiplication table

×	22	22	22	22	22	22	22	22
♟	22	22	22	22	22	22	22	22
2	44	44	44	44	44	44	44	44
♗	66	66	66	66	66	66	66	66
4	88	88	88	88	88	88	88	88
♖	110	110	110	110	110	110	110	110
6	132	132	132	132	132	132	132	132
7	154	154	154	154	154	154	154	154
8	176	176	176	176	176	176	176	176
♕	198	198	198	198	198	198	198	198

Multiplication table

×	33	33	33	33	33	33	33	33
♙	33	33	33	33	33	33	33	33
2	66	66	66	66	66	66	66	66
♗	99	99	99	99	99	99	99	99
4	132	132	132	132	132	132	132	132
♖	165	165	165	165	165	165	165	165
6	198	198	198	198	198	198	198	198
7	231	231	231	231	231	231	231	231
8	264	264	264	264	264	264	264	264
♕	297	297	297	297	297	297	297	297

Ho Math Chess 何数棋谜 妈！我会棋谜式乘法啦！
Mom! I Learn Multiplication Using Math-Chess-Puzzles Connection!
Student's Name _____ Date _____
2007 - 2017 © Frank Ho, Amanda Ho, All rights reserved.   www.homathchess.com

Multiplication table

×	44	44	44	44	44	44	44	44
♙	44	44	44	44	44	44	44	44
2	88	88	88	88	88	88	88	88
♗	132	132	132	132	132	132	132	132
4	176	176	176	176	176	176	176	176
♖	220	220	220	220	220	220	220	220
6	264	264	264	264	264	264	264	264
7	308	308	308	308	308	308	308	308
8	352	352	352	352	352	352	352	352
♕	396	396	396	396	396	396	396	396

Ho Math Chess   何数棋谜   妈！我会棋谜式乘法啦！
Mom! I Learn Multiplication Using Math-Chess-Puzzles Connection!

Student's Name _____ Date _____

2007 - 2017 © Frank Ho, Amanda Ho, All rights reserved.   www.homathchess.com

Multiplication table

✕	55	55	55	55	55	55	55	55
♙	55	55	55	55	55	55	55	55
2	110	110	110	110	110	110	110	110
♗	165	165	165	165	165	165	165	165
4	220	220	220	220	220	220	220	220
♖	275	275	275	275	275	275	275	275
6	330	330	330	330	330	330	330	330
7	385	385	385	385	385	385	385	385
8	440	440	440	440	440	440	440	440
♕	495	495	495	495	495	495	495	495

Ho Math Chess 何数棋谜 妈！我会棋谜式乘法啦！
Mom! I Learn Multiplication Using Math-Chess-Puzzles Connection!
Student's Name _____ Date _____

2007 - 2017 © Frank Ho, Amanda Ho, All rights reserved. www.homathchess.com

Multiplication table

×	66	66	66	66	66	66	66	66
♙	66	66	66	66	66	66	66	66
2	132	132	132	132	132	132	132	132
♗	198	198	198	198	198	198	198	198
4	264	264	264	264	264	264	264	264
♖	330	330	330	330	330	330	330	330
6	396	396	396	396	396	396	396	396
7	462	462	462	462	462	462	462	462
8	528	528	528	528	528	528	528	528
♕	594	594	594	594	594	594	594	594

2007 - 2017 © Frank Ho, Amanda Ho, All rights reserved.   www.homathchess.com

Multiplication table

×	77	77	77	77	77	77	77	77
♙	77	77	77	77	77	77	77	77
2	154	154	154	154	154	154	154	154
♗	231	231	231	231	231	231	231	231
4	308	308	308	308	308	308	308	308
♖	385	385	385	385	385	385	385	385
6	462	462	462	462	462	462	462	462
7	539	539	539	539	539	539	539	539
8	616	616	616	616	616	616	616	616
♕	693	693	693	693	693	693	693	693

Ho Math Chess 何数棋谜 妈！我会棋谜式乘法啦！
Mom! I Learn Multiplication Using Math-Chess-Puzzles Connection!
Student's Name _____ Date _____

2007 - 2017 © Frank Ho, Amanda Ho, All rights reserved. www.homathchess.com

Multiplication table

×	88	88	88	88	88	88	88	88
♟	88	88	88	88	88	88	88	88
2	176	176	176	176	176	176	176	176
♗	264	264	264	264	264	264	264	264
4	352	352	352	352	352	352	352	352
♖	440	440	440	440	440	440	440	440
6	528	528	528	528	528	528	528	528
7	616	616	616	616	616	616	616	616
8	704	704	704	704	704	704	704	704
♕	792	792	792	792	792	792	792	792

Ho Math Chess   何数棋谜   妈！我会棋谜式乘法啦！
Mom! I Learn Multiplication Using Math-Chess-Puzzles Connection!

Student's Name _____ Date _____

2007 - 2017 © Frank Ho, Amanda Ho, All rights reserved.   www.homathchess.com

Multiplication table

×	99	99	99	99	99	99	99	99
♙	99	99	99	99	99	99	99	99
2	198	198	198	198	198	198	198	198
♗	297	297	297	297	297	297	297	297
4	396	396	396	396	396	396	396	396
♖	495	495	495	495	495	495	495	495
6	594	594	594	594	594	594	594	594
7	693	693	693	693	693	693	693	693
8	792	792	792	792	792	792	792	792
♕	891	891	891	891	891	891	891	891

2007 - 2017 © Frank Ho, Amanda Ho, All rights reserved.　　www.homathchess.com

ddd ✕ d with carrying

$$
\begin{array}{r} 112 \\ \times\ 6 \\ \hline 672 \end{array}
\qquad
\begin{array}{r} 112 \\ \times\ 9 \\ \hline 1008 \end{array}
\qquad
\begin{array}{r} 112 \\ \times\ 7 \\ \hline 784 \end{array}
\qquad
\begin{array}{r} 112 \\ \times\ 8 \\ \hline 896 \end{array}
\qquad
\begin{array}{r} 112 \\ \times\ ♛ \\ \hline 1008 \end{array}
$$

$$
\begin{array}{r} 113 \\ \times\ ♜ \\ \hline 565 \end{array}
\qquad
\begin{array}{r} 113 \\ \times\ 6 \\ \hline 678 \end{array}
\qquad
\begin{array}{r} 113 \\ \times\ 7 \\ \hline 791 \end{array}
\qquad
\begin{array}{r} 113 \\ \times\ 8 \\ \hline 904 \end{array}
\qquad
\begin{array}{r} 113 \\ \times\ 9 \\ \hline 1017 \end{array}
$$

$$
\begin{array}{r} 114 \\ \times\ 6 \\ \hline 684 \end{array}
\qquad
\begin{array}{r} 114 \\ \times\ ♛ \\ \hline 1026 \end{array}
\qquad
\begin{array}{r} 114 \\ \times\ 7 \\ \hline 798 \end{array}
\qquad
\begin{array}{r} 114 \\ \times\ 8 \\ \hline 912 \end{array}
\qquad
\begin{array}{r} 114 \\ \times\ ♜ \\ \hline 570 \end{array}
$$

$$
\begin{array}{r} 115 \\ \times\ ♜ \\ \hline 575 \end{array}
\qquad
\begin{array}{r} 115 \\ \times\ 6 \\ \hline 690 \end{array}
\qquad
\begin{array}{r} 115 \\ \times\ ♛ \\ \hline 1035 \end{array}
\qquad
\begin{array}{r} 115 \\ \times\ 8 \\ \hline 920 \end{array}
\qquad
\begin{array}{r} 115 \\ \times\ 7 \\ \hline 805 \end{array}
$$

Ho Math Chess   何数棋谜   妈！我会棋谜式乘法啦！
Mom! I Learn Multiplication Using Math-Chess-Puzzles Connection!

Student's Name _____ Date _____

2007 - 2017 © Frank Ho, Amanda Ho, All rights reserved.   www.homathchess.com

ddd × d with carrying

2 2 2	2 2 2	2 2 2	2 2 2	2 2 2
x 6	x 9	x 7	x 8	X ♖
1332	1998	1554	1776	1110

2 2 3	2 2 3	2 2 3	2 2 3	2 2 3
x 6	X ♕	x 7	x 8	x 5
1338	2007	1561	1784	1115

2 2 4	2 2 4	2 2 4	2 2 4	2 2 4
x 6	X ♖	x 7	x 8	x 5
1344	1120	1568	1792	1120

2 2 5	2 2 5	2 2 5	2 2 5	2 2 5
X ♖	x 9	x 7	X ♕	x 5
1125	1025	1575	2025	1125

2007 - 2017 © Frank Ho, Amanda Ho, All rights reserved.　www.homathchess.com

ddd × d with carrying

3 3 2	3 3 2	3 3 2	3 3 2	3 3 2
x　　6	x　♛	x　　7	x　　8	x　♜
1992	2988	2324	2656	1660
3 3 3	3 3 3	3 3 3	3 3 3	3 3 3
x　　6	x　♛	x　　7	x　　8	x　　5
1998	2997	2331	2664	1665
3 3 4	3 3 4	3 3 4	3 3 4	3 3 4
x　　6	x　♜	x　　7	x　　8	x　　5
2004	1670	2338	2672	1670
3 3 5	3 3 5	3 3 5	3 3 5	3 3 5
x　　6	x　　9	x　♜	x　　8	x　♛
2010	3015	1675	2680	3015

2007 - 2017 © Frank Ho, Amanda Ho, All rights reserved.   www.homathchess.com

ddd × d with carrying

442	442	442	442	442
x 6	x 9	x 7	x 8	x 5
2652	3978	3094	3536	2210

443	443	443	443	443
x 6	x ♛	x 7	x 8	x ♜
2658	3987	3101	3544	2215

444	444	444	444	444
x 6	x ♛	x 7	x 8	x 5
2664	3996	3108	3552	2220

445	445	445	445	445
x 6	x 8	x 7	x ♛	x ♜
2670	3560	3115	4005	2225

Ho Math Chess　何数棋谜　妈！我会棋谜式乘法啦！
Mom! I Learn Multiplication Using Math-Chess-Puzzles Connection!

Student's Name _____ Date _____

2007 - 2017 © Frank Ho, Amanda Ho, All rights reserved.　　www.homathchess.com

ddd × d with carrying

5 5 2	5 5 2	5 5 2	5 5 2	5 5 2
x　6	x　8	x　7	x　♛	x　5
3112	4416	3864	4968	2760

5 5 3	5 5 3	5 5 3	5 5 3	5 5 3
x　♜	x　8	x　7	x　♛	x　6
2765	4424	3871	4977	3318

5 5 4	5 5 4	5 5 4	5 5 4	5 5 4
x　6	x　9	x　♜	x　8	x　7
3324	4986	2770	4432	3878

5 5 5	5 5 5	5 5 5	5 5 5	5 5 5
x　6	x　♛	x　7	x　8	x　♜
3330	4995	3885	4440	2775

ddd × d with carrying

6 6 2	6 6 2	6 6 2	6 6 2	6 6 2
x 6	x ♛	x 7	x 8	x 5
3972	5958	4634	5296	3310

6 6 3	6 6 3	6 6 3	6 6 3	6 6 3
x 6	x 8	x 7	x ♛	x ♜
3978	5304	4641	5967	3315

6 6 4	6 6 4	6 6 4	6 6 4	6 6 4
x 6	x 9	x 7	x 8	x ♜
3984	5976	4648	5312	3320

6 6 5	6 6 5	6 6 5	6 6 5	6 6 5
x 6	x ♜	x 7	x 8	x ♛
3990	3325	4655	5320	5985

2007 - 2017 © Frank Ho, Amanda Ho, All rights reserved.   www.homathchess.com

ddd × d with carrying

```
  772      772      772      772      772
x    6   x    9   x    7   x    8   x    ▦
 4632     6948     5404     6176     3860
```

```
  773      773      773      773      773
x    6   x    ♛   x    7   x    8   x    5
 4638     6957     5411     6184     3865
```

```
  774      774      774      774      774
x    6   x    9   x    7   x    8   x    ▦
 4644     6966     5418     6192     3870
```

```
  775      775      775      775      775
x    6   x    ♛   x    7   x    8   x    5
 4650     6975     5425     6200     3875
```

Ho Math Chess 何数棋谜 妈！我会棋谜式乘法啦！
Mom! I Learn Multiplication Using Math-Chess-Puzzles Connection!

Student's Name _____ Date _____

2007 - 2017 © Frank Ho, Amanda Ho, All rights reserved.   www.homathchess.com

ddd ✕ d with carrying

8 8 2	8 8 2	8 8 2	8 8 2	8 8 2
✕ 6	✕ ♜	✕ 7	✕ 8	✕ ♛
5292	4410	6174	7056	7938

8 8 3	8 8 3	8 8 3	8 8 3	8 8 3
✕ 6	✕ ♛	✕ 7	✕ ♜	✕ 8
5298	7947	6181	4415	7064

8 8 4	8 8 4	8 8 4	8 8 4	8 8 4
✕ ♜	✕ 9	✕ 7	✕ 8	✕ 6
4420	7956	6188	7072	5304

8 8 5	8 8 5	8 8 5	8 8 5	8 8 5
✕ 6	✕ ♛	✕ 7	✕ 8	✕ ♜
5310	7965	6195	7080	4425

Mom! I Learn Multiplication Using Math-Chess-Puzzles Connection!

Student's Name _____ Date _____

2007 - 2017 © Frank Ho, Amanda Ho, All rights reserved.   www.homathchess.com

ddd × d with carrying

9 9 2	9 9 2	9 9 9	9 9 2	9 9 2
x 6	X ♛	X ♜	x 8	x 7
5952	8928	4995	7936	6944

9 9 3	9 9 3	9 9 3	9 9 3	9 9 3
x 6	X ♜	x 7	x 8	X ♛
5958	4965	6951	7944	8937

9 9 4	9 9 4	9 9 4	9 9 4	9 9 4
x 6	x 8	x 7	X ♛	X ♜
5964	7952	6958	8946	4970

9 9 5	9 9 5	9 9 5	9 9 5	9 9 5
x 6	X ♛	X ♜	x 8	x 7
5970	8955	4975	7960	6965

Ho Math Chess   何数棋谜 妈！我会棋谜式乘法啦！
Mom! I Learn Multiplication Using Math-Chess-Puzzles Connection!

Student's Name _____ Date _____

2007 - 2017 © Frank Ho, Amanda Ho, All rights reserved.   www.homathchess.com

Changing the order of d X dd to dd X d

Comparison of 3 X 1234 is changed to the offer of 1234 X 3.

Regular way as presented	Bigger number X smaller number

Ho Math Chess   何数棋谜   妈！我会棋谜式乘法啦！
Mom! I Learn Multiplication Using Math-Chess-Puzzles Connection!

Student's Name _____ Date _____

2007 - 2017 © Frank Ho, Amanda Ho, All rights reserved.   www.homathchess.com

dd X d0s (X by multiples of 10's, equivalent to ÷ multiples of 0.1)

Place the number ending with 0s as the second factor and just bring down 0s.

$$\begin{array}{r} 21 \\ \times\ 20 \\ \hline \square\square \\ \square\square \\ \hline \square\square\square \\ 420 \end{array}$$

$$\begin{array}{r} 21 \\ \times\ \ 20 \\ \hline \square\square 0 \\ 420 \end{array}$$

Just bring down one zero.

$$\begin{array}{r} 19 \\ \times\ 200 \\ \hline \square\square \\ \square\square \\ \square\square \\ \hline \square\square\square\square \\ 3800 \end{array}$$

$$\begin{array}{r} 19 \\ \times\ \ 200 \\ \hline \square\square 00 \\ 3800 \end{array}$$

Just bring down two zeros.

2 X 1000 = 2000   999 X 1000 = 999000

300 X 1000 = 300 000   909 X 1000 = 909000

2100 X 100 = 210000   101 X 1000 = 101000

99 X 1000 = 99000   123 X 1000 = 123000

10000 X 540 = 5400000   10000 X 111 = 1110000

2000, 300000, 210000, 99000, 5400000, 999000, 909000
101000, 123000, 1110000

2007 - 2017 © Frank Ho, Amanda Ho, All rights reserved.　　www.homathchess.com

dd X d0s (X by multiples of 10's, equivalent to ÷ multiples of 0.1)

Place the number ending with 0s as the second factor and just bring down 0s.

300 X 123 = 123 X 300?

Regular way	Bring down zeros.
3 0 0 X 1 2 3 □□□ □□□ □□□ □□□□□ 36900	1 2 3 X 3 0 0 □□□□□ 36900

36900, 36900

dd X d0s (X by multiples of 10's, equivalent to ÷ multiples of 0.1)

Place the number ending with 0s as the second factor and just bring down 0s.

123 X 12000 1470000	1 2 3 X 1 2 0 0 0 1476000
13000 X 14 182000	
234 X 12000 2808000	

1476000, 182000, 2808000

2007 - 2017 © Frank Ho, Amanda Ho, All rights reserved.   www.homathchess.com

dd × **d0**

$$
\begin{array}{r} 28 \\ \times\ 20 \\ \hline 560 \end{array}
\qquad
\begin{array}{r} 28 \\ \times\ 20 \\ \hline 560 \end{array}
\qquad
\begin{array}{r} 23 \\ \times\ 50 \\ \hline 1150 \end{array}
\qquad
\begin{array}{r} 23 \\ \times\ 50 \\ \hline 1150 \end{array}
$$

$$
\begin{array}{r} 46 \\ \times\ 20 \\ \hline 920 \end{array}
\qquad
\begin{array}{r} 39 \\ \times\ 60 \\ \hline 2340 \end{array}
\qquad
\begin{array}{r} 79 \\ \times\ 50 \\ \hline 3950 \end{array}
\qquad
\begin{array}{r} 17 \\ \times\ 40 \\ \hline 680 \end{array}
$$

$$
\begin{array}{r} 26 \\ \times\ 80 \\ \hline 2080 \end{array}
\qquad
\begin{array}{r} 53 \\ \times\ 30 \\ \hline 1590 \end{array}
\qquad
\begin{array}{r} 47 \\ \times\ 60 \\ \hline 2820 \end{array}
\qquad
\begin{array}{r} 78 \\ \times\ 70 \\ \hline 5460 \end{array}
$$

Ho Math Chess 何数棋谜 妈！我会棋谜式乘法啦！
Mom! I Learn Multiplication Using Math-Chess-Puzzles Connection!

Student's Name _____ Date _____

2007 - 2017 © Frank Ho, Amanda Ho, All rights reserved.   www.homathchess.com

dd × d0

$$
\begin{array}{r} 43 \\ \times\ 30 \\ \hline 1290 \end{array}
\qquad
\begin{array}{r} 53 \\ \times\ 40 \\ \hline 2120 \end{array}
\qquad
\begin{array}{r} 63 \\ \times\ 30 \\ \hline 1890 \end{array}
\qquad
\begin{array}{r} 27 \\ \times\ 90 \\ \hline 2430 \end{array}
$$

$$
\begin{array}{r} 76 \\ \times\ 70 \\ \hline 5320 \end{array}
\qquad
\begin{array}{r} 58 \\ \times\ 60 \\ \hline 3480 \end{array}
\qquad
\begin{array}{r} 45 \\ \times\ 50 \\ \hline 2250 \end{array}
\qquad
\begin{array}{r} 53 \\ \times\ 80 \\ \hline 4240 \end{array}
$$

$$
\begin{array}{r} 87 \\ \times\ 90 \\ \hline 7830 \end{array}
\qquad
\begin{array}{r} 39 \\ \times\ 80 \\ \hline 3120 \end{array}
\qquad
\begin{array}{r} 95 \\ \times\ 50 \\ \hline 4750 \end{array}
\qquad
\begin{array}{r} 67 \\ \times\ 40 \\ \hline 2680 \end{array}
$$

dd ✕ d0

12 X 20 = 240　　　　32 X 30 =960　　　　52 X 50 =2600

33 X 40 = 1320　　　52 X 60 =3120　　　63 X 20 =1260

32 X 70 = 2240　　　54 X 40 =2160　　　17 X 80 =1360

27 X 90 = 2430　　　62 X 60 =3720　　　72 X 30 =2160

dd × d0s

27	29	75	96
x 4 0 0	x 5 0 0	x 7 0 0	x 4 0 0
10800	14500	52500	38400

1 9	7 3	4 9	3 5
x 3 0 0	x 4 0 0	x 6 0 0	x 9 0 0
5700	29200	29400	31500

9 1	5 3	7 6	8 3
x 8 0 0	x 7 0 0	x 5 0 0	x 9 0 0
72800	37100	38000	74700

2007 - 2017 © Frank Ho, Amanda Ho, All rights reserved. www.homathchess.com

dd $\times$ d0s

42 X 600 = 25200   37 X 200 =7400   98 X 400 =39200

28 X 700 = 19600   48 X 300 =14400   43 X 400 =17200

17 X 900 = 15300   27 X 500 =13500   61 X 200 =12200

24 X 900 = 21600   73 X 500 =36500   34 X 400 =13600

2007 - 2017 © Frank Ho, Amanda Ho, All rights reserved.　　www.homathchess.com

dd $\times$ dd multiplication concepts

Horizontal multiplication

$23 \times 24 = 23 \times (4 + 20) = 23 \times 4 + 23 \times 20 = 92 + 460 = 552$

Vertical multiplication

It is extremely important for teacher to explain the concepts of why and how multiplication is done. Repeated drills without explanations only add prolong learning curve which could have been reduced if concepts such as the following had been explained.

1. Why line up ones multiplication at the rightmost position? Because it is ones place.
2. Why a 0 is placed at the ones place when doing tens place multiplication? Because tens place multiplication always has a 0 at ones place such as 20 in the example, the ones place value is 0.
3. How is the horizontal multiplication related to the vertical multiplication? The vertical multiplication is using the concept of distributive law to do the work but written in a vertical way.

2007 - 2017 © Frank Ho, Amanda Ho, All rights reserved.　　www.homathchess.com

Step 1	Step 2
$$\begin{array}{r} 2\ 3 \\ \times\quad 4 \\ \hline 9\ 2 \end{array}$$	$$\begin{array}{r} 2\ 3 \\ \times\quad 2\ 0 \\ \hline 4\ 6\ 0 \end{array}$$

Step 3: The answer is 92 + 460 = 552

$25 \times 24 = 25 \times (4 + 20) = 25 \times 4 + 25 \times 20 = 600$

$24 \times 25 = 24 \times (5 + 20) = 24 \times 5 + 24 \times 20 = 600$

$25 \times 36 = 25 \times (6 + 30) = 25 \times 6 + 25 \times 30 = 900$

$36 \times 25 = 36 \times (5 + 20) = 36 \times 5 + 36 \times 20 = 900$

$27 \times 28 = 27 \times (8 + 20) = 27 \times 8 + 27 \times 20 = 756$

$28 \times 27 = 28 \times (7 + 20) = 28 \times 7 + 28 \times 20 = 756$

600, 600, 900, 900, 756, 756

dd × dd with carrying

1	2	1	2	2
12	21	23	24	23
× 16	× 16	× 16	× 16	× 17
72	126	138	144	161
12	21	23	24	23
192	336	368	384	391

$$15 \times 15$$
□
□
□□
□□□
225

$$16 \times 27$$
□□□
□□
□□□
432

$$17 \times 28$$
□□□
□□
□□□
476

$$18 \times 29$$
□□□
□□
□□□
522

dd × dd with carrying

```
    15          15          15              15
  x 16        x  17        x  18          x  19
    □        □□□        □□□          □□□
    □          □□          □□            □□
   □□

  □□□        □□□        □□□          □□□
   240         255         270             285
```

```
    15          18          19              21
  x 16        x  17        x  18          x  19
    □        □□□        □□□          □□□
    □          □□          □□            □□
   □□

  □□□        □□□        □□□          □□□
   240         306         342             399
```

2007 - 2017 © Frank Ho, Amanda Ho, All rights reserved.   www.homathchess.com

dd × dd with carrying

```
    2 5          2 6          2 7          2 8
  x 1 6        x 1 7        x 1 9        x 2 1
  □ □ □        □ □ □        □ □ □        □ □ □
    □ □          □ □          □ □          □ □
  □ □ □        □ □ □        □ □ □        □ □ □
  4 0 0        4 4 2        5 1 3        5 8 8
```

```
    2 6          2 7          2 8          2 9
  x 1 6        x 2 7        x 2 8        x 2 9
  □ □ □        □ □ □        □ □ □        □ □ □
    □ □          □ □          □ □          □ □
  □ □ □        □ □ □        □ □ □        □ □ □
  4 1 6        7 2 9        7 8 4        8 4 1
```

2007 - 2017 © Frank Ho, Amanda Ho, All rights reserved.   www.homathchess.com

dd × dd with carrying

```
    37          38          39              49
  x 16        x  17       x  19          x  21
  □□□         □□□         □□□             □□
   □□          □□          □□             □□
  ─────       ─────       ─────          ──────
  □□□         □□□         □□□            □□□□
   592         646         741           1029


    48          49          51              51
  x 16        x  17       x  18          x  19
  □□□         □□□         □□□             □□□
   □□          □□          □□              □□
  ─────       ─────       ─────          ─────
  □□□         □□□         □□□             □□□
   768         833         918            969
```

dd × dd with carrying

	2 9		2 9		2 9		2 9
x	1 6	x	1 7	x	1 8	x	1 9

□□□   □□□   □□□   □□□
□□   □□   □□   □□

□□□   □□□   □□□   □□□
4 6 4   4 9 3   5 2 2   5 5 1

	3 5		3 5		3 5		3 5
x	1 6	x	1 7	x	1 8	x	1 9

□□□   □□□   □□□   □□□
□□   □□   □□   □□

□□□   □□□   □□□   □□□
5 6 0   5 9 5   6 3 0   6 6 5

dd × dd with carrying

```
   36          36          36           36
 x 16        x 17        x 18         x 19
 □□□        □□□        □□□          □□□
  □□          □□          □□            □□
 ───         ───         ───           ───
 □□□        □□□        □□□          □□□
 576         612         648           684
```

```
   37          37          37           37
 x 16        x 17        x 18         x 19
 □□□        □□□        □□□          □□□
  □□          □□          □□            □□
 ───         ───         ───           ───
 □□□        □□□        □□□          □□□
 592         629         666           703
```

576, 612, 648, 684
592, 629, 666, 703

Ho Math Chess 何数棋谜 妈！我会棋谜式乘法啦！
Mom! I Learn Multiplication Using Math-Chess-Puzzles Connection!

Student's Name _____ Date _____

2007 - 2017 © Frank Ho, Amanda Ho, All rights reserved.   www.homathchess.com

dd × dd with carrying

```
    38          39              46              56
  x 26        x 27           x  28           x  29
  □□□         □□□             □□□             □□□
  □□          □□              □□              □□□
  □□□         □□□□            □□□□            □□□□
   988        1053            1288            1624


    57          58              59              67
  x 26        x 27           x  28           x  29
  □□□         □□□             □□              □□□
  □□□         □□□             □               □□□
                              □□□
  □□□□        □□□□            □□□□            □□□□
  1482        1566            1652            1943
```

988, 1053, 1288, 1624
1482, 1566, 1652, 1943

Ho Math Chess   何数棋谜   妈！我会棋谜式乘法啦！
Mom! I Learn Multiplication Using Math-Chess-Puzzles Connection!

Student's Name _____ Date _____

2007 - 2017 © Frank Ho, Amanda Ho, All rights reserved.   www.homathchess.com

dd × dd with carrying

```
      1           2           1           2
   1 2         2 1         2 3         2 4
 x 1 6       x 1 6       x 1 6       x 1 6
 ─────       ─────       ─────       ─────
   7 2       1 2 6       1 3 8       1 4 4
 1 2         2 1         2 3         2 4
 ─────       ─────       ─────       ─────
 1 9 2       3 3 6       3 6 8       3 8 4
```

```
   1 5         1 6         1 7         1 8
 x 1 5       x   2 7     x   2 8     x   2 9
 ─────       ─────────   ─────────   ─────────
    □         □ □ □       □ □ □       □ □ □
    □           □ □         □ □         □ □
  □ □
 ─────       ─────────   ─────────   ─────────
 □ □ □       □ □ □       □ □ □       □ □ □
 2 2 5       4 3 2       4 7 6       5 2 2
```

2007 - 2017 © Frank Ho, Amanda Ho, All rights reserved.　　www.homathchess.com

dd × dd with carrying

$$
\begin{array}{r} 15 \\ \times\ 16 \\ \hline \end{array}
\qquad
\begin{array}{r} 15 \\ \times\ 17 \\ \hline \end{array}
\qquad
\begin{array}{r} 15 \\ \times\ 18 \\ \hline \end{array}
\qquad
\begin{array}{r} 15 \\ \times\ 19 \\ \hline \end{array}
$$

240　　255　　270　　285

$$
\begin{array}{r} 15 \\ \times\ 16 \\ \hline \end{array}
\qquad
\begin{array}{r} 18 \\ \times\ 17 \\ \hline \end{array}
\qquad
\begin{array}{r} 19 \\ \times\ 18 \\ \hline \end{array}
\qquad
\begin{array}{r} 21 \\ \times\ 19 \\ \hline \end{array}
$$

240　　306　　342　　399

240, 255, 270, 285
240, 306, 342, 399

2007 - 2017 © Frank Ho, Amanda Ho, All rights reserved.   www.homathchess.com

dd ✕ dd without carrying

47	32	23	38
x 11	x 22	x 33	x 11
517	704	759	418

42	32	29	37
x 22	x 22	x 11	x 11
924	704	319	407

27	33	37	14
x 11	x 22	x 11	x 22
297	726	407	308

Ho Math Chess 何数棋谜 妈！我会棋谜式乘法啦！

Mom! I Learn Multiplication Using Math-Chess-Puzzles Connection!

Student's Name _____ Date _____

2007 - 2017 © Frank Ho, Amanda Ho, All rights reserved. www.homathchess.com

dd ✕ dd with carrying

$$
\begin{array}{r} 45 \\ \times\ 25 \\ \hline 1125 \end{array}
\qquad
\begin{array}{r} 36 \\ \times\ 46 \\ \hline 1656 \end{array}
\qquad
\begin{array}{r} 27 \\ \times\ 57 \\ \hline 1539 \end{array}
\qquad
\begin{array}{r} 38 \\ \times\ 48 \\ \hline 1824 \end{array}
$$

$$
\begin{array}{r} 49 \\ \times\ 29 \\ \hline 1421 \end{array}
\qquad
\begin{array}{r} 37 \\ \times\ 47 \\ \hline 1739 \end{array}
\qquad
\begin{array}{r} 25 \\ \times\ 55 \\ \hline 1375 \end{array}
\qquad
\begin{array}{r} 37 \\ \times\ 47 \\ \hline 1739 \end{array}
$$

$$
\begin{array}{r} 33 \\ \times\ 22 \\ \hline 726 \end{array}
\qquad
\begin{array}{r} 22 \\ \times\ 33 \\ \hline 726 \end{array}
\qquad
\begin{array}{r} 24 \\ \times\ 22 \\ \hline 528 \end{array}
\qquad
\begin{array}{r} 44 \\ \times\ 22 \\ \hline 968 \end{array}
$$

Ho Math Chess　何数棋谜　妈！我会棋谜式乘法啦！
Mom! I Learn Multiplication Using Math-Chess-Puzzles Connection!

Student's Name _____ Date _____

2007 - 2017 © Frank Ho, Amanda Ho, All rights reserved.　　www.homathchess.com

dd ✕ **dd with carrying**

27	36	37	18
x 27	x 56	x 67	x 78
729	2016	2479	1404

59	26	27	88
x 29	x 46	x 37	x 68
1711	1196	999	5984

17	26	25	48
x 21	x 42	x 56	x 36
357	1092	1400	1728

Ho Math Chess　何数棋谜　妈！我会棋谜式乘法啦！

Mom! I Learn Multiplication Using Math-Chess-Puzzles Connection!

Student's Name _____ Date _____

2007 - 2017 © Frank Ho, Amanda Ho, All rights reserved.　www.homathchess.com

dd ✕ dd with carrying

$$
\begin{array}{r} 45 \\ \times\ 19 \\ \hline 855 \end{array}
\qquad
\begin{array}{r} 12 \\ \times\ 47 \\ \hline 564 \end{array}
\qquad
\begin{array}{r} 29 \\ \times\ 55 \\ \hline 1595 \end{array}
\qquad
\begin{array}{r} 37 \\ \times\ 23 \\ \hline 851 \end{array}
$$

$$
\begin{array}{r} 27 \\ \times\ 11 \\ \hline 297 \end{array}
\qquad
\begin{array}{r} 11 \\ \times\ 56 \\ \hline 616 \end{array}
\qquad
\begin{array}{r} 57 \\ \times\ 26 \\ \hline 1482 \end{array}
\qquad
\begin{array}{r} 14 \\ \times\ 52 \\ \hline 728 \end{array}
$$

$$
\begin{array}{r} 32 \\ \times\ 21 \\ \hline 672 \end{array}
\qquad
\begin{array}{r} 23 \\ \times\ 42 \\ \hline 966 \end{array}
\qquad
\begin{array}{r} 24 \\ \times\ 32 \\ \hline 768 \end{array}
\qquad
\begin{array}{r} 24 \\ \times\ 13 \\ \hline 312 \end{array}
$$

dd ✕ dd with carrying

47	36	27	38
x 28	x 46	x 56	x 48
1316	1656	1512	1824

45	32	29	37
x 29	x 47	x 55	x 43
1305	1504	1595	1591

27	33	37	18
x 29	x 56	x 66	x 72
783	1848	2442	1296

Ho Math Chess 何数棋谜 妈！我会棋谜式乘法啦！

Mom! I Learn Multiplication Using Math-Chess-Puzzles Connection!

Student's Name _____ Date _____

2007 - 2017 © Frank Ho, Amanda Ho, All rights reserved.   www.homathchess.com

dd ✕ dd with carrying

$$\begin{array}{r} 57 \\ \times\,29 \\ \hline 1653 \end{array} \qquad \begin{array}{r} 26 \\ \times\,46 \\ \hline 1196 \end{array} \qquad \begin{array}{r} 27 \\ \times\,36 \\ \hline 972 \end{array} \qquad \begin{array}{r} 88 \\ \times\,66 \\ \hline 5808 \end{array}$$

$$\begin{array}{r} 51 \\ \times\,28 \\ \hline 1428 \end{array} \qquad \begin{array}{r} 36 \\ \times\,42 \\ \hline 1512 \end{array} \qquad \begin{array}{r} 29 \\ \times\,31 \\ \hline 899 \end{array} \qquad \begin{array}{r} 68 \\ \times\,62 \\ \hline 4216 \end{array}$$

$$\begin{array}{r} 63 \\ \times\,84 \\ \hline 5292 \end{array} \qquad \begin{array}{r} 78 \\ \times\,46 \\ \hline 3588 \end{array} \qquad \begin{array}{r} 84 \\ \times\,25 \\ \hline 2100 \end{array} \qquad \begin{array}{r} 74 \\ \times\,68 \\ \hline 5032 \end{array}$$

Ho Math Chess　　何数棋谜　妈！我会棋谜式乘法啦！
Mom! I Learn Multiplication Using Math-Chess-Puzzles Connection!

Student's Name _____ Date _____

2007 - 2017 © Frank Ho, Amanda Ho, All rights reserved.　　www.homathchess.com

dd ✕ dd with carrying

$$
\begin{array}{r} 47 \\ \times\,28 \\ \hline 1316 \end{array}
\qquad
\begin{array}{r} 36 \\ \times\,46 \\ \hline 1656 \end{array}
\qquad
\begin{array}{r} 27 \\ \times\,56 \\ \hline 1512 \end{array}
\qquad
\begin{array}{r} 38 \\ \times\,46 \\ \hline 1748 \end{array}
$$

$$
\begin{array}{r} 45 \\ \times\,29 \\ \hline 1305 \end{array}
\qquad
\begin{array}{r} 32 \\ \times\,47 \\ \hline 1504 \end{array}
\qquad
\begin{array}{r} 29 \\ \times\,55 \\ \hline 1595 \end{array}
\qquad
\begin{array}{r} 37 \\ \times\,43 \\ \hline 1591 \end{array}
$$

$$
\begin{array}{r} 27 \\ \times\,29 \\ \hline 783 \end{array}
\qquad
\begin{array}{r} 33 \\ \times\,56 \\ \hline 1848 \end{array}
\qquad
\begin{array}{r} 37 \\ \times\,66 \\ \hline 2442 \end{array}
\qquad
\begin{array}{r} 18 \\ \times\,72 \\ \hline 1296 \end{array}
$$

Mom! I Learn Multiplication Using Math-Chess-Puzzles Connection!

Student's Name _____ Date _____

2007 - 2017 © Frank Ho, Amanda Ho, All rights reserved.   www.homathchess.com

dd ✕ **dd with carrying**

12 X 28 = 336   32 X 27 =864   52 X 35 =1820

33 X 45 = 1485   52 X 25 =1300   63 X 21 =1323

32 X 27 = 864   54 X 18 =972   17 X 48 =816

27 X 83 = 2241   62 X 24 =1488   72 X 32 =2304

Ho Math Chess   何数棋谜   妈！我会棋谜式乘法啦！

Mom! I Learn Multiplication Using Math-Chess-Puzzles Connection!

Student's Name _____ Date _____

2007 - 2017 © Frank Ho, Amanda Ho, All rights reserved.   www.homathchess.com

dd ✕ dd with carrying

$42 \times 38 = 1596$ $35 \times 25 = 875$ $52 \times 25 = 1300$

$38 \times 45 = 1710$ $48 \times 25 = 1200$ $69 \times 21 = 1449$

$76 \times 27 = 2052$ $36 \times 41 = 1476$ $95 \times 28 = 2660$

$62 \times 73 = 4526$ $53 \times 26 = 1378$ $74 \times 36 = 2664$

2007 - 2017 © Frank Ho, Amanda Ho, All rights reserved.　　www.homathchess.com

dd ✕ dd with carrying

54 X 23 = 1242　　　62 X 23 =1426　　　71 X 85 =6035

53 X 45 = 2385　　　82 X 75 =6150　　　63 X 48 =3024

81 X 37 = 2997　　　76 X 52 =3952　　　49 X 68 =3332

77 X 84 = 6468　　　93 X 54 =5022　　　82 X 39 =3198

2007 - 2017 © Frank Ho, Amanda Ho, All rights reserved.   www.homathchess.com

dd $\times$ 10s

$35 \times 10 = 350$ $57 \times 10 = 570$ $26 \times 10 = 260$

$53 \times 100 = 5300$ $21 \times 100 = 2100$ $64 \times 100 = 6400$

$84 \times 1000 = 84000$ $86 \times 1000 = 86000$ $24 \times 1000 = 24000$

$64 \times 10 = 640$ $57 \times 100 = 5700$ $97 \times 1000 = 97000$

2007 - 2017 © Frank Ho, Amanda Ho, All rights reserved.   www.homathchess.com

dd $\times$ **5**

For advanced students only.

42 X 5
=21 X 2 X 5
=21 X 10
=210

38 X 5 =190

52 X 5 =260

36 X 5 = 180

48 X 5 =240

64 X 5 =320

76 X 5 = 380

56 X 5 =280

96 X 5 =480

62 X 5 = 310

54 X 5 =270

74 X 5 =370

dd $\times$ **25**

For advanced students only.

8 X 25
=2 X 4 X 25
=2 X 100
=200

36 X 25 =900

48 X 25 =1200

24 X 25 = 600

20 X 25 =500

16 X 25 =400

12 X 25 = 300

32 X 25 =800

28 X 25 =700

44 X 25 = 1100

40 X 25 =1000

56 X 25 =1400

Ho Math Chess 何数棋谜 妈！我会棋谜式乘法啦！
Mom! I Learn Multiplication Using Math-Chess-Puzzles Connection!

Student's Name _____ Date _____
2007 - 2017 © Frank Ho, Amanda Ho, All rights reserved. www.homathchess.com

dd × 11

23 X 11
=23 X (10+1)
=23 X 10+23 X 1
=230+23
=253

32 X 11 =352

52 X 11 =572

31 X 11 = 341

45 X 11 =495

44 X 11 =484

71 X 11 = 781

56 X 11 =616

96 X 11 =1056

68 X 11 = 748

57 X 11 =627

74 X 11 =814

2007 - 2017 © Frank Ho, Amanda Ho, All rights reserved.　　www.homathchess.com

ddd ✕ dd with carrying

```
     125              126              127
 x    16          x    17          x    15
   □ □ □            □ □ □            □ □ □
  □ □ □            □ □ □            □ □ □
 □ □ □ □          □ □ □ □          □ □ □ □
  2 0 0 0          2 1 4 2          1 9 0 5
```

```
     234              235              236
 x    56          x    57          x    58
  □ □ □ □          □ □ □ □          □ □ □ □
  □ □ □ □          □ □ □ □          □ □ □ □
 □ □ □ □ □        □ □ □ □ □        □ □ □ □ □
 1 3 1 0 4        1 3 3 9 5        1 3 6 8 8
```

2000, 2142, 1905
13104, 13395, 13688

2007 - 2017 © Frank Ho, Amanda Ho, All rights reserved.   www.homathchess.com

ddd × dd with carrying

```
    345              345              345
  x   65           x   67           x   68
  ┌─┬─┬─┬─┐       ┌─┬─┬─┬─┐       ┌─┬─┬─┬─┐
  └─┴─┴─┴─┘       └─┴─┴─┴─┘       └─┴─┴─┴─┘
  ┌─┬─┬─┬─┐       ┌─┬─┬─┬─┐       ┌─┬─┬─┬─┐
  └─┴─┴─┴─┘       └─┴─┴─┴─┘       └─┴─┴─┴─┘
  ─────────       ─────────       ─────────
  ┌─┬─┬─┬─┬─┐     ┌─┬─┬─┬─┬─┐     ┌─┬─┬─┬─┬─┐
  └─┴─┴─┴─┴─┘     └─┴─┴─┴─┴─┘     └─┴─┴─┴─┴─┘
   2 2 4 2 5       2 3 1 1 5       2 3 4 6 0


    666              777              888
  x   66           x   77           x   88
  ┌─┬─┬─┬─┐       ┌─┬─┬─┬─┐       ┌─┬─┬─┬─┐
  └─┴─┴─┴─┘       └─┴─┴─┴─┘       └─┴─┴─┴─┘
  ┌─┬─┬─┬─┐       ┌─┬─┬─┬─┐       ┌─┬─┬─┬─┐
  └─┴─┴─┴─┘       └─┴─┴─┴─┘       └─┴─┴─┴─┘
  ─────────       ─────────       ─────────
  ┌─┬─┬─┬─┬─┐     ┌─┬─┬─┬─┬─┐     ┌─┬─┬─┬─┬─┐
  └─┴─┴─┴─┴─┘     └─┴─┴─┴─┴─┘     └─┴─┴─┴─┴─┘
   4 3 9 5 6       5 9 8 2 9       7 8 1 4 4
```

2007 - 2017 © Frank Ho, Amanda Ho, All rights reserved.   www.homathchess.com

ddd × dd with carrying

$$
\begin{array}{r}
345 \\
\times \quad 77 \\
\hline
\square\square\square\square \\
\square\square\square\square \\
\hline
\square\square\square\square\square \\
26565
\end{array}
\qquad
\begin{array}{r}
345 \\
\times \quad 88 \\
\hline
\square\square\square\square \\
\square\square\square\square \\
\hline
\square\square\square\square\square \\
30360
\end{array}
\qquad
\begin{array}{r}
345 \\
\times \quad 99 \\
\hline
\square\square\square\square \\
\square\square\square\square \\
\hline
\square\square\square\square\square \\
34155
\end{array}
$$

$$
\begin{array}{r}
999 \\
\times \quad 77 \\
\hline
\square\square\square\square \\
\square\square\square\square \\
\hline
\square\square\square\square\square \\
76923
\end{array}
\qquad
\begin{array}{r}
999 \\
\times \quad 88 \\
\hline
\square\square\square\square \\
\square\square\square\square \\
\hline
\square\square\square\square\square \\
87912
\end{array}
\qquad
\begin{array}{r}
999 \\
\times \quad 99 \\
\hline
\square\square\square\square \\
\square\square\square\square \\
\hline
\square\square\square\square\square \\
98901
\end{array}
$$

Ho Math Chess   何数棋谜   妈!我会棋谜式乘法啦!
Mom! I Learn Multiplication Using Math-Chess-Puzzles Connection!

Student's Name _____ Date _____

2007 - 2017 © Frank Ho, Amanda Ho, All rights reserved.   www.homathchess.com

ddd × dd with carrying

$$\begin{array}{r} 567 \\ \times\ \ 77 \\ \hline \square\square\square\square \\ \square\square\square\square \\ \hline \square\square\square\square\square \\ 4\ 3\ 6\ 5\ 9 \end{array}$$

$$\begin{array}{r} 765 \\ \times\ \ 88 \\ \hline \square\square\square\square \\ \square\square\square\square \\ \hline \square\square\square\square\square \\ 6\ 7\ 3\ 2\ 0 \end{array}$$

$$\begin{array}{r} 675 \\ \times\ \ 99 \\ \hline \square\square\square\square \\ \square\square\square\square \\ \hline \square\square\square\square\square \\ 6\ 6\ 8\ 2\ 5 \end{array}$$

$$\begin{array}{r} 678 \\ \times\ \ 77 \\ \hline \square\square\square\square \\ \square\square\square\square \\ \hline \square\square\square\square\square \\ 5\ 2\ 2\ 0\ 6 \end{array}$$

$$\begin{array}{r} 876 \\ \times\ \ 88 \\ \hline \square\square\square\square \\ \square\square\square\square \\ \hline \square\square\square\square\square \\ 7\ 7\ 0\ 8\ 8 \end{array}$$

$$\begin{array}{r} 786 \\ \times\ \ 99 \\ \hline \square\square\square\square \\ \square\square\square\square \\ \hline \square\square\square\square\square \\ 7\ 7\ 8\ 1\ 4 \end{array}$$

2007 - 2017 © Frank Ho, Amanda Ho, All rights reserved.   www.homathchess.com

ddd ✕ dd without carrying

```
   472        425         432         384
 x  11      x  22       x  33       x  11
  5192       9350       14256        4224


   323        421         224         352
 x  22      x  22       x  11       x  11
  7106       9262        2464        3872


   286        412         374         243
 x  11      x  22       x  11       x  22
  3146       9064        4114        5346
```

2007 - 2017 © Frank Ho, Amanda Ho, All rights reserved.   www.homathchess.com

ddd ✕ dd with carrying

474	316	237	589
x 15	x 36	x 28	x 31
7110	11376	6636	18259

457	628	275	874
x 52	x 27	x 61	x 39
23764	16956	16775	34086

577	368	779	147
x 61	x 42	x 43	x 29
35197	15456	33497	4263

Ho Math Chess   何数棋谜   妈！我会棋谜式乘法啦！
Mom! I Learn Multiplication Using Math-Chess-Puzzles Connection!

Student's Name _____ Date _____

2007 - 2017 © Frank Ho, Amanda Ho, All rights reserved.   www.homathchess.com

ddd $\times$ dd with carrying

159 X 64 = 10176   357 X 24 =8568   684 X 86 =58824

268 X 57 = 15276   842 X 17 =14314   368 X 43 =15824

251 X 74 = 18574   961 X 19 =18259   254 X 24 =6096

741 X 27 = 20007   217 X 14 =3038   364 X 43 =15652

2007 - 2017 © Frank Ho, Amanda Ho, All rights reserved.   www.homathchess.com

ddd ✕ dd with carrying

148 X 24 = 3552   572 X 47 =26884   351 X 34 =11934

381 X 56 = 21336   729 X 72 =52488   349 X 764=266636

941 X 48 = 45168   149 X 24 =3576   324 X 27 =8748

327 X 34 = 11118   243 X 63 =15309   751 X 82 =61582

d0d × **dd**

305 X 47 = 14335　　507 X 55 =27885　　805 X 51 =41055

609 X 34 = 20706　　709 X 78 =55302　　501 X 39 =19539

207 X 49 = 10143　　608 X 56 =34048　　901 X 27 =24327

206 X 91 = 18746　　604 X 28 =16912　　908 X 62 =56296

2007 - 2017 © Frank Ho, Amanda Ho, All rights reserved.   www.homathchess.com

dd X d0d

The 0s in the middle of a factor do not change the sum, so no product needs to be done.

Long form

```
    1 2 3
x   1 0 1
  ☐☐☐
 ☐☐☐
☐☐☐
─────────
☐☐☐☐☐
 1 2 4 2 3
```

Short form

```
    1 2 3
x   1 0 1
  ☐☐☐
(Do not multiply 0)
  ☐☐☐
─────────
☐☐☐☐☐
 1 2 4 2 3
```

```
  2 1
x 2 0 1
  ☐☐
 ☐☐
─────
☐☐☐☐
4 2 2 1
```

```
  2 1
x 3 0 1
  ☐☐
 ☐☐
─────
☐☐☐☐
6 3 2 1
```

```
  1 9
x 4 0 1
  ☐☐
 ☐☐
─────
☐☐☐☐
7 6 1 9
```

```
  1 9
x 5 0 1
  ☐☐
 ☐☐
─────
☐☐☐☐
9 5 1 9
```

12423, 12423, 4221, 6321, 7619, 9519

Student's Name _____ Date _____

2007 - 2017 © Frank Ho, Amanda Ho, All rights reserved.   www.homathchess.com

d0d ✕ dd

```
    603        708        503        805
  x  23      x  37      x  48      x  51
  1 8 0 9     26196      24144      41055
1 2 0 6
1 3 8 69


    704        209        307        406
  x  72      x  87      x  67      x  35
  50688      18183      20569      14210


    504        604        209        807
  x  51      x  72      x  93      x  49
  25704      43488      19437      39543
```

2007 - 2017 © Frank Ho, Amanda Ho, All rights reserved.　　www.homathchess.com

d0d × dd

```
    502        407        705        903
  x   37     x   42     x   51     x   45
   18574      17094      35955      40635


    205        706        408        105
  x   17     x   17     x   38     x   18
    3485      12002      15504       1890


    802        708        204        809
  x   43     x   71     x   86     x   16
   34486      50268      17544      12944
```

2007 - 2017 © Frank Ho, Amanda Ho, All rights reserved.　www.homathchess.com

d0d $\times$ **dd**

609 X 47 = 28623　　304 X 24 =7296　　605 X 38 =22990

508 X 57 = 28956　　702 X 45 =31590　　308 X 29 =8932

801 X 43 = 34443　　501 X 24 =12024　　804 X 37 =29748

401 X 52 = 20852　　807 X 41 =33087　　204 X 62 =12648

dd X d0d

The 0s in the middle of a factor do not change the sum, so no product needs to be done.

```
    2 0 1        2 0 1        1 0 9        1 0 9
x   2 0 1    x   3 0 1    x   4 0 1    x   5 0 1
    □ □ □        □ □ □        □ □ □        □ □ □
  □ □ □ □      □ □ □ □      □ □ □ □      □ □ □ □
 □ □ □ □ □    □ □ □ □ □    □ □ □ □ □    □ □ □ □ □
  4 0 4 0 1    6 0 5 0 1    4 3 7 0 9    5 4 6 0 9
```

```
    2 0 0 1        2 0 0 1        3 0 0 1        4 0 0 1
x   2 0 0 1    x   3 0 0 1    x   4 0 0 1    x   5 0 0 1
   4004001        6005001       12007001       20009001
```

40401, 60601, 43709, 54609
4004001, 16006001, 12007001, 20009001

2007 - 2017 © Frank Ho, Amanda Ho, All rights reserved.   www.homathchess.com

1s X 1s

11 X 11 = 121   111 X 11 =1221   1111 X 11 =12221

11111 X 11 =   111111 X 11   1111111 X 11
122221   =1222221   =12222221

111 X 111 = 12321   1111 X 1111   11111 X 11111
                     =1234321   =123454321

222 X 111 = 24642   333 X 111 =36963   444 X 111 =49284

2007 - 2017 © Frank Ho, Amanda Ho, All rights reserved.　　www.homathchess.com

1s X 1s

101 X 5 = 505   101 X 8 =808   101 X 6 =606

101 X 11 = 1111   101 X 26 =2626   101 X 75 =7575

101 X 111 = 11211   101 X 123 =12423   101 X 532 =53732

101 X 45 = 4545   101 X 78 =7878   101 X 642 =64842

2007 - 2017 © Frank Ho, Amanda Ho, All rights reserved.   www.homathchess.com

5s X 5s

$15 \times 15 = 225$ $25 \times 25 = 625$ $35 \times 35 = 1225$

$45 \times 45 = 2025$ $55 \times 55 = 3025$ $65 \times 65 = 4225$

$75 \times 75 = 5625$ $85 \times 85 = 7225$ $95 \times 95 = 9025$

$25 \times 25 = 625$ $45 \times 45 = 2025$ $75 \times 75 = 5625$

d × d × d

8 X 6 X 9
= 48X 9
=432

11 X 25 X 9
=2475

13 X 14 X 8
=1456

21 X 15 X 34
=10710

32 X 11 X 72
=25344

17 X 18 X 35
=10710

24 X 32 X 17
=13056

27 X 31 X 52
=43524

33 X 47 X 41
=63591

52 X 13 X 26
=17576

43 X 82 X 24
=84624

48 X 41 X 24
=47232

2007 - 2017 © Frank Ho, Amanda Ho, All rights reserved.　www.homathchess.com

Replace the ? with a number.

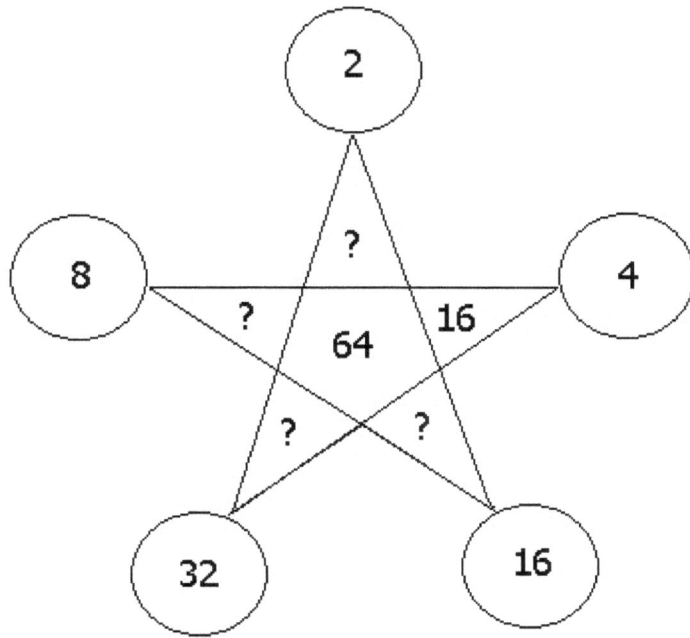

32， 16， 4， 2， 8

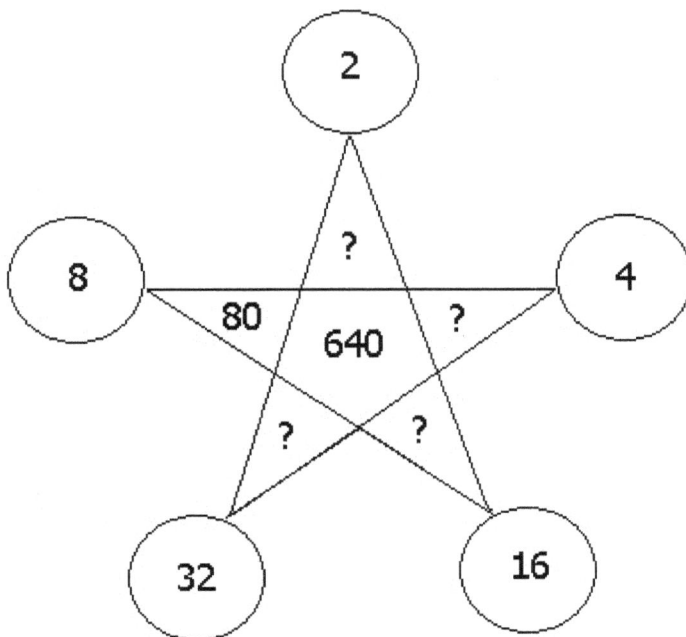

320， 160， 40， 20， 80

2007 - 2017 © Frank Ho, Amanda Ho, All rights reserved.　　www.homathchess.com

Replace the ? with a number.

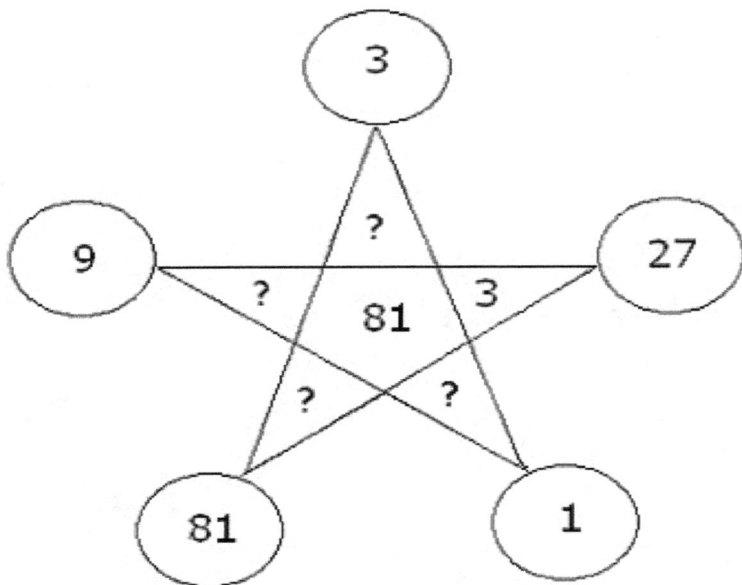

27，3，81，1，9

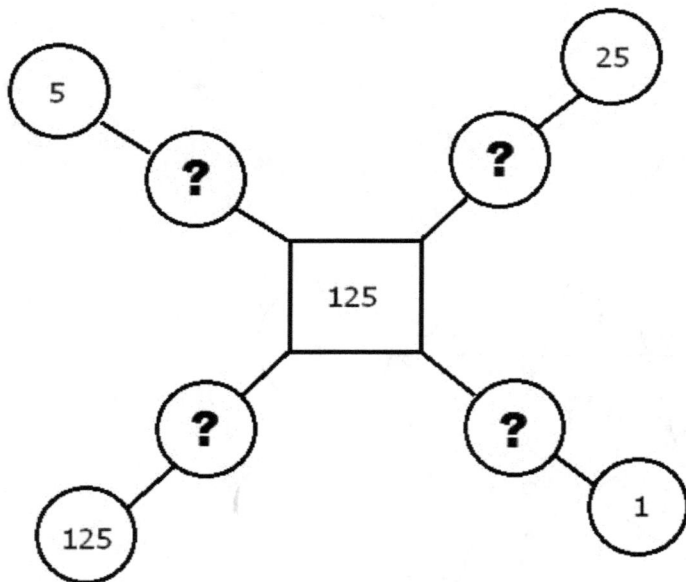

25，5，125，1

2007 - 2017 © Frank Ho, Amanda Ho, All rights reserved.　　www.homathchess.com

Power of 10

$5 \times 2 = 10$

$25 \times 4 = 100$

$125 \times 8 = 1000$

The above strategy can be used in multiplication or used in converting fraction to percent.

$5 \times 2 \times 5 \times 2 \times 5 \times 2 = 1\ 000$

$25 \times 4 \times 25 \times 4 \times 25 = 250\ 000$

$125 \times 8 \times 125 \times 8 \times 125 \times 8 = 1\ 000\ 000\ 000$

$5 \times 2 \times 5 \times 2 \times 5 \times 2 \times 5 \times 2 = 10\ 000$

$25 \times 4 \times 25 \times 4 \times 25 \times 4 \times 25 \times 4 \times = 100\ 000\ 000$

$125 \times 8 \times 125 \times 8 \times 125 \times 8 \times 125 \times 8 = 1\ 000\ 000\ 000\ 000$

From multiplication to division procedure

In this workbook, the division procedure is based on the reverse procedure of multiplication, For example, $\square \times 2 = 6$, the division procedure is to find out what is the greatest factor of $\square$ such that $\square \times 2 \leq 6$ with the following division notation

$$2\overline{)6}$$

There are a few restrictions to follow such as the divisor must be a whole number and the remainder must be less than the divisor and divide the dividends one digit at a time. The reason of dividing the dividend digit one at a time is it automatically takes care of the zeros at the beginning, in the middle, and at the end. The divisor must be whole number would make the decimal division easier.

A simple problem could be used to demonstrate the above concept. For example, $25 is to be divided equally among 5 children. The divisor is 5 children so it must be whole number, but the $ amount could be decimals. If we would write the division as $25 \div 5 = \frac{\$25}{5}$ then

$$\frac{\$25}{5} = \frac{2\$10 + 1\$5}{5} = 2\$2 + \$1 = \$5$$

If we follow the above procedure for division, it would be very tedious, so we convert 2 $10 to 20 of $1 and add $5 to be 5 of $1. So the division procedure will be

$$5\overline{)25}$$. This is the reason that quotient 5 must be placed rightmost to show we are

actually divide 25 of 1$ by 5 children.

For decimal division, the division flowchart included in this workbook can also be used. Bring down 0 until the desired decimal places are found. If the dividend has a decimal point, line up decimal point in the quotient and just carry out division as it is whole number division. If the dividend is a whole number, place decimal point at the place after the dividend digit has been all used and before 0 needs to be brought down.

2007 - 2017 © Frank Ho, Amanda Ho, All rights reserved.   www.homathchess.com

From multiplication to division (d ÷ d)

Multiplication	Division
factor $\times$ factor $\leq$ product	$divisor\sqrt{dividend}$ with $quotinet$ above
$3\square \times 2 \leq 6$	$\times\ \square 3\leftarrow$ Think what times 2 is ≤ 6. $2\overline{)6}$ $-\ \square \leftarrow$ Do subtraction $0\leftarrow$ Remainder (what is left) = 0
$4\square \times 2 \leq 8$	$\times\ \square 4\leftarrow$ Think what times 2 is ≤ 8. $2\overline{)8}$ $-\ \square \leftarrow$ Do subtraction $0\leftarrow$ Remainder (what is left) = 0
$3\square \times 3 \leq$ ♛	$\times\ \square 3\leftarrow$ Think what times 2 is ≤ 9. $3\overline{)9}$ $-\ \square \leftarrow$ Do subtraction $0\leftarrow$ Remainder (what is left) = 0

Ho Math Chess　何数棋谜　妈！我会棋谜式乘法啦！
Mom! I Learn Multiplication Using Math-Chess-Puzzles Connection!
Student's Name _____ Date _____
2007 - 2017 © Frank Ho, Amanda Ho, All rights reserved.　　www.homathchess.com

From multiplication to division (d ÷ d)

$4\Box \times 2 \leq 8$	$8 = \Box 4 \times 2$	$2\overline{)8}$ ← step 1: What times 2 is ≤ 8. step 2: $8 - 8 = 0$　0 ← Remainder = 0
$0\Box \times 3 \leq 0$	$0 = \Box 0 \times ♗$	$3\overline{)0}$ ← step 1: What times 3 is ≤ 0. step 2: $0 - 0 = 0$　0 ← Remainder = 0
$1\Box \times 6 \leq 6$	$6 = \Box 1 \times 6$	$6\overline{)6}$ ← step 1: What times 6 is ≤ 6. step 2: $6 - 6 = 0$　0 ← Remainder = 0

2007 - 2017 © Frank Ho, Amanda Ho, All rights reserved.   www.homathchess.com

dd ÷ d with 1-digit quotient and no remainder

$7\square \times 6 = 42$	$42 = \square_6 \times 7$	Step1: Do $\square \times 6 \le 42$ (4 is too small, use 42) **Place the quotient in the rightmost position** $\times\ \square_7 \leftarrow$ step 2: $7 \times 6 = 42$ $6\overline{)42}$ $-\square\square \leftarrow$ step 3: $42 - 42 = 0$ $0 \leftarrow$ Remainder $= 0$
$9\square \times 2 = 18$	$18 = \square_2 \times ♛$	**Step1:** Do $\square \times 2 \le 18$ (1 is too small, use 18) **Place the quotient in the rightmost position** $\times\ \square_9 \leftarrow$ step 2: $9 \times 2 = 18$ $2\overline{)18}$ $-\square\square \leftarrow$ step 3: $18 - 18 = 0$ $0 \leftarrow$ Remainder $= 0$
$9\square \times ♛ = 81$	$81 = \square_9 \times 9$	**Step1:** Do $\square \times 9 \le 81$ (8 is too small, use 81) **Place the quotient in the rightmost position** $\times\ \square_9 \leftarrow$ step 2: $9 \times 9 = 81$ $9\overline{)81}$ $-\square\square \leftarrow$ step 3: $81 - 81 = 0$ $0 \leftarrow$ Remainder $= 0$

2007 - 2017 © Frank Ho, Amanda Ho, All rights reserved.   www.homathchess.com

dd ÷ d with 1-digit quotinet and no remainder

Step1: Do ☐ ✕ 3 ≤ 15 (1 is too small, use 15)

✕ ☐5 ← step 2: Do multiplication, 5 ✕ 3 =

15

3)15

− ☐☐ ← step 3: Do subtraction, 15 − 15 = 0

0 ←Remainder = 0

Step1 Do ☐ ✕ 3 ≤ 18 (3 is too small, use 18)

✕ ☐6 ← step 2: Do multiplication, 6 ✕ 3 = 18

3)18

− ☐☐ ← step 3: Do substraction,18 − 18 =

0

0 ←Remainder = 0

Step1: Do ☐ ✕ 5 ≤ 25 (2 is too small, use 25)

✕ ☐5 ← step 2: Do multiplication

5)25

− ☐☐ ← step 3: Do subtraction

0 ←Remainder = 0

Step1: Do ☐ ✕ 4 ≤ 28 (2 is too small, use 28)

✕ ☐7 ← step 2: Do multiplication

4)28

− ☐☐ ← step 3: Do subtraction

0 ←Remainder = 0

Step1: Do ☐ ✕ 4 ≤ 32 (3 is too small, use 32)

✕ ☐8 ← step 2: Do multiplication

4)32

− ☐☐ ← step 3: Do subtraction

0 ←Remainder = 0

Step1: Do ☐ ✕ 2 ≤ 16 (1 is too small, use 16)

✕ ☐8 ← step 2: Do multiplication

2)16

− ☐☐ ← step 3: Do subtraction

0 ←Remainder = 0

From multiplication to division

$6\square \times 2 \leq 12$ $2\square \times 6 \leq 12$	$^{\times}\square 2$ $6\overline{)12}$ $\square\square$	$^{\times}\square 6$ $2\overline{)12}$ $\square\square$	$12 \div 2 = \square 6$ $12 \div 6 = \square 2$
$6\square \times 3 \leq 18$ $3\square \times \ \leq 18$	$^{\times}\square 3$ $6\overline{)18}$ $\square\square$	$^{\times}\square 6$ $3\overline{)18}$ $\square\square$	$18 \div ♞ = \square 6$ $18 \div 6 = \square 3$
$3\square \times 4 \leq 12$ $4\square \times 3 \leq 12$	$^{\times}\square 3$ $4\overline{)12}$ $\square\square$	$^{\times}\square 4$ $3\overline{)12}$ $\square\square$	$12 \div 4 = \square 3$ $12 \div ♗ = \square 4$

Student's Name _____ Date _____

2007 - 2017 © Frank Ho, Amanda Ho, All rights reserved.   www.homathchess.com

From multiplication to division

$3\Box \times 5 \leq 15$ $5\Box \times 3 \leq 15$	$^x\Box 3$ $5\overline{)15}$ $\Box\Box$	$^x\Box 5$ $3\overline{)15}$ $\Box\Box$	$15 \div ♗ = \Box 5$ $15 \div ♖ = \Box 3$
$6\Box \times ♖ \leq 30$ $5\Box \times 6 \leq 30$	$^x\Box 5$ $6\overline{)30}$ $\Box\Box$	$^x\Box 6$ $5\overline{)30}$ $\Box\Box$	$30 \div 5 = \Box 6$ $30 \div 6 = \Box 5$
$8\Box \times 5 \leq 40$ $5\Box \times 8 \leq 40$	$^x\Box 8$ $5\overline{)40}$ $\Box\Box$	$^x\Box 5$ $8\overline{)40}$ $\Box\Box$	$40 \div 5 = \Box 8$ $40 \div 8 = \Box 5$

2007 - 2017 © Frank Ho, Amanda Ho, All rights reserved.　　www.homathchess.com

From multiplication to division

$9\square \times 5 \le 45$ $5\square \times 9 \le 45$	$\begin{array}{r} ^{\times}\square\,9 \\ 5\overline{)45} \\ \square\square \end{array}$	$\begin{array}{r} ^{\times}\square\,5 \\ 9\overline{)45} \\ \square\square \end{array}$	$45 \div 9 = \square\,5$ $45 \div ♜ = \square\,9$
$9\square \times ♜ \le 10$ $5\square \times 2 \le 10$	$\begin{array}{r} ^{\times}\square\,5 \\ 2\overline{)10} \\ \square\square \end{array}$	$\begin{array}{r} ^{\times}\square\,2 \\ 5\overline{)10} \\ \square\square \end{array}$	$10 \div 5 = \square\,2$ $10 \div 2 = \square\,5$
$5\square \times 5 \le 25$ $5\square \times ♜ \le 25$	$\begin{array}{r} ^{\times}\square\,5 \\ 5\overline{)25} \\ \square\square \end{array}$	$\begin{array}{r} ^{\times}\square\,5 \\ 5\overline{)25} \\ \square\square \end{array}$	$25 \div 5 = \square\,5$ $25 \div 5 = \square\,5$

From multiplication to division

$5\square \times 4 \le 20$ $4\square \times 5 \le 20$	$\times \square 4$ $5\overline{)20}$ $\square\square$	$\times \square 5$ $4\overline{)20}$ $\square\square$	$20 \div 4 = \square 5$ $20 \div \text{♖} = \square 4$
$7\square \times \text{♖} \le 35$ $5\square \times 7 \le 35$	$\times \square 5$ $7\overline{)35}$ $\square\square$	$\times \square 7$ $5\overline{)35}$ $\square\square$	$35 \div 5 = \square 7$ $35 \div 7 = \square 5$
$5\square \times 8 \le 40$ $8\square \times \text{♖} \le 40$	$\times \square 8$ $5\overline{)40}$ $\square\square$	$\times \square 5$ $8\overline{)40}$ $\square\square$	$40 \div \text{♖} = \square 8$ $40 \div 8 = \square 5$

2007 - 2017 © Frank Ho, Amanda Ho, All rights reserved. www.homathchess.com

From multiplication to division (d ÷ d)

$6\square \times 4 \leq 24$ $4\square \times 6 \leq 24$	$\times \square 6$ $4\overline{)24}$ $\square\square$	$\times \square 4$ $6\overline{)24}$ $\square\square$	$24 \div 4 = \square 6$ $24 \div 6 = \square 4$
$7\square \times 6 \leq 42$ $6\square \times 7 \leq 42$	$\times \square 6$ $7\overline{)42}$ $\square\square$	$\times \square 7$ $6\overline{)42}$ $\square\square$	$42 \div 6 = \square 7$ $42 \div 7 = \square 6$
$6\square \times 6 \leq 36$ $6\square \times 6 \leq 36$	$\times \square 6$ $6\overline{)36}$ $\square\square$	$\times \square 6$ $6\overline{)36}$ $\square\square$	$36 \div 6 = \square 6$ $36 \div 6 = \square 6$

Ho Math Chess   何数棋谜   妈！我会棋谜式乘法啦！
Mom! I Learn Multiplication Using Math-Chess-Puzzles Connection!

Student's Name _____ Date _____

2007 - 2017 © Frank Ho, Amanda Ho, All rights reserved.   www.homathchess.com

dd ÷ d with 1-digit quotient and no remainder

$6\square \times 5 \le 30$ $6\square \times ♖ \le 30$	$^{\times}\square 5$ $6\overline{)30}$ $\square\square$	$^{\times}\square 6$ $5\overline{)30}$ $\square\square$	$30 \div 5 = \square 6$ $30 \div 6 = \square 5$
$4\square \times 6 \le 24$ $6\square \times 4 \le 24$	$^{\times}\square 6$ $4\overline{)24}$ $\square\square$	$^{\times}\square 4$ $6\overline{)24}$ $\square\square$	$24 \div 6 = \square 4$ $24 \div 4 = \square 6$
$6\square \times ♗ \le 18$ $3\square \times 6 \le 18$	$^{\times}\square 3$ $6\overline{)18}$ $\square\square$	$^{\times}\square 6$ $3\overline{)18}$ $\square\square$	$18 \div 6 = \square 3$ $18 \div ♗ = \square 6$

2007 - 2017 © Frank Ho, Amanda Ho, All rights reserved.　　www.homathchess.com

dd ÷ d with 1-digit quotient and no remainder

$_9\square \times 5 \le 45$ $_5\square \times ♛ \le 45$	$\begin{array}{r} x\,\square\,9 \\ 5\overline{)45} \\ \square\square \end{array}$	$\begin{array}{r} x\,\square\,5 \\ 9\overline{)45} \\ \square\square \end{array}$	$45 \div 5 = \square_9$ $45 \div ♛ = \square_5$
$_8\square \times ♖ \le 40$ $_5\square \times 8 \le 40$	$\begin{array}{r} x\,\square\,8 \\ 5\overline{)40} \\ \square\square \end{array}$	$\begin{array}{r} x\,\square\,5 \\ 8\overline{)40} \\ \square\square \end{array}$	$40 \div 5 = \square_8$ $40 \div 8 = \square_5$
$_5\square \times 6 \le 30$ $_6\square \times 5 \le 30$	$\begin{array}{r} x\,\square\,5 \\ 6\overline{)30} \\ \square\square \end{array}$	$\begin{array}{r} x\,\square\,6 \\ 5\overline{)30} \\ \square\square \end{array}$	$30 \div 6 = \square_5$ $30 \div ♖ = \square_6$

Student's Name _____ Date _____

2007 - 2017 © Frank Ho, Amanda Ho, All rights reserved.   www.homathchess.com

dd ÷ d with 1-digit quotient and no remainder

$7\square \times ♖ \le 35$ $5\square \times 7 \le 35$	$\times \square 7$ $5\overline{)35}$ $\square\square$	$\times \square 5$ $7\overline{)35}$ $\square\square$	$35 \div ♖ = \square 7$ $35 \div 7 = \square 5$
$6\square \times 5 \le 30$ $5\square \times 6 \le 30$	$\times \square 6$ $5\overline{)30}$ $\square\square$	$\times \square 5$ $6\overline{)30}$ $\square\square$	$30 \div ♖ = \square 6$ $30 \div 6 = \square 5$
$5\square \times 4 \le 20$ $4\square \times ♖ \le 20$	$\times \square 4$ $5\overline{)20}$ $\square\square$	$\times \square 5$ $4\overline{)20}$ $\square\square$	$20 \div 4 = \square 5$ $20 \div ♖ = \square 4$

Ho Math Chess   何数棋谜 妈！我会棋谜式乘法啦！
Mom! I Learn Multiplication Using Math-Chess-Puzzles Connection!

Student's Name _____ Date _____

2007 - 2017 © Frank Ho, Amanda Ho, All rights reserved.   www.homathchess.com

dd ÷ d with remainder vs. no remainder

$35 \div 7 = \boxed{}\ {}_5$	$\begin{array}{r} {}^{\times \square 7} \\ 5\overline{)35} \\ \square\square \end{array}$	$\begin{array}{r} {}^{\times \square 7} \\ 5\overline{)36} \\ \square\square \\ \square \end{array}$
$20 \div \text{♜} = \boxed{}\ {}_4$	$\begin{array}{r} {}^{\times \square 5} \\ 4\overline{)20} \\ \square\square \end{array}$	$\begin{array}{r} {}^{\times \square 5} \\ 4\overline{)22} \\ \square\square \\ \square \end{array}$
$56 \div 8 = \boxed{}\ {}_7$	$\begin{array}{r} {}^{\times \square 8} \\ 7\overline{)56} \\ \square\square \end{array}$	$\begin{array}{r} {}^{\times \square 8} \\ 7\overline{)59} \\ \square\square \\ \square \end{array}$
$81 \div \text{♛} = \boxed{}\ {}_9$	$\begin{array}{r} {}^{\times \square 9} \\ 9\overline{)81} \\ \square\square \end{array}$	$\begin{array}{r} {}^{\times \square 9} \\ 9\overline{)85} \\ \square\square \\ \square \end{array}$

2007 - 2017 © Frank Ho, Amanda Ho, All rights reserved.   www.homathchess.com

dd ÷ d with remainder vs. no remainder

$21 \div \text{♘} = \square\,7$	$7)\overline{21}$ ×□3 □□	$7)\overline{25}$ ×□3 □□ □4
$42 \div 6 = \square\,7$	$7)\overline{42}$ ×□6 □□	$7)\overline{45}$ ×□6 □□ □3
$32 \div 4 = \square\,8$	$8)\overline{32}$ ×□4 □□	$8)\overline{38}$ ×□4 □□ □6
$30 \div 6 = \square\,5$	$5)\overline{30}$ ×□6 □□	$5)\overline{33}$ ×□6 □□ □3

dd ÷ d with remainder vs. no remainder

$18 \div ♛ = \square\,2$	$x\,\square9$ $2\,)\,\overline{18}$ $\square\square$	$x\,\square9$ $2\,)\,\overline{19}$ $\square\square$ $\square1$
$12 \div 6 = \square\,2$	$x\,\square6$ $2\,)\,\overline{12}$ $\square\square$	$x\,\square6$ $2\,)\,\overline{13}$ $\square\square$ $\square1$
$14 \div 7 = \square\,2$	$x\,\square7$ $2\,)\,\overline{14}$ $\square\square$	$x\,\square7$ $2\,)\,\overline{15}$ $\square\square$ $\square1$
$28 \div 4 = \square\,7$	$x\,\square4$ $7\,)\,\overline{28}$ $\square\square$	$x\,\square4$ $7\,)\,\overline{33}$ $\square\square$ $\square5$

dd ÷ d with remainder vs. no remainder

$15 \div$ ♖ $= \square\,3$	$\begin{array}{r} ^{\times}\square 5 \\ 3\overline{)15} \\ \square\square \end{array}$	$\begin{array}{r} ^{\times}\square 5 \\ 3\overline{)16} \\ \square\square \\ \square 1 \end{array}$
$18 \div 6 = \square\,3$	$\begin{array}{r} ^{\times}\square 6 \\ 3\overline{)18} \\ \square\square \end{array}$	$\begin{array}{r} ^{\times}\square 6 \\ 3\overline{)19} \\ \square\square \\ \square 1 \end{array}$
$24 \div 8 = \square\,3$	$\begin{array}{r} ^{\times}\square 8 \\ 3\overline{)24} \\ \square\square \end{array}$	$\begin{array}{r} ^{\times}\square 8 \\ 3\overline{)26} \\ \square\square \\ \square 2 \end{array}$
$27 \div$ ♛ $= \square\,3$	$\begin{array}{r} ^{\times}\square 9 \\ 3\overline{)27} \\ \square\square \end{array}$	$\begin{array}{r} ^{\times}\square 9 \\ 3\overline{)29} \\ \square\square \\ \square 2 \end{array}$

Ho Math Chess 何数棋谜 妈！我会棋谜式乘法啦！
Mom! I Learn Multiplication Using Math-Chess-Puzzles Connection!

Student's Name _____ Date _____

2007 - 2017 © Frank Ho, Amanda Ho, All rights reserved.   www.homathchess.com

dd ÷ d with remainder vs. no remainder

$12 \div ♗ = \square\,4$	$\begin{array}{r} {}^{\times}\square 3 \\ 4\overline{)12} \\ \square\square \end{array}$	$\begin{array}{r} {}^{\times}\square 3 \\ 4\overline{)13} \\ \square\square \\ \square 1 \end{array}$
$16 \div 4 = \square\,4$	$\begin{array}{r} {}^{\times}\square 4 \\ 4\overline{)16} \\ \square\square \end{array}$	$\begin{array}{r} {}^{\times}\square 4 \\ 4\overline{)18} \\ \square\square \\ \square 2 \end{array}$
$20 \div ♖ = \square\,4$	$\begin{array}{r} {}^{\times}\square 5 \\ 4\overline{)20} \\ \square\square \end{array}$	$\begin{array}{r} {}^{\times}\square 5 \\ 4\overline{)23} \\ \square\square \\ \square 3 \end{array}$
$24 \div 6 = \square\,4$	$\begin{array}{r} {}^{\times}\square 6 \\ 4\overline{)24} \\ \square\square \end{array}$	$\begin{array}{r} {}^{\times}\square 6 \\ 4\overline{)27} \\ \square\square \\ \square 3 \end{array}$

2007 - 2017 © Frank Ho, Amanda Ho, All rights reserved.   www.homathchess.com

dd ÷ d with remainder vs. no remainder

$10 \div 2 = \square\,5$	$5\overline{)10}$ $^\times\square 2$ $\square\square$	$5\overline{)14}$ $^\times\square 2$ $\square\square$ $\square 4$
$20 \div 4 = \square\,5$	$5\overline{)20}$ $^\times\square 4$ $\square\square$	$5\overline{)24}$ $^\times\square 4$ $\square\square$ $\square 4$
$25 \div ♖ = \square\,5$	$5\overline{)25}$ $^\times\square 5$ $\square\square$	$5\overline{)24}$ $^\times\square 4$ $\square\square$ $\square 4$
$40 \div 8 = \square\,5$	$5\overline{)40}$ $^\times\square 8$ $\square\square$	$5\overline{)44}$ $^\times\square 8$ $\square\square$ $\square 4$

2007 - 2017 © Frank Ho, Amanda Ho, All rights reserved.   www.homathchess.com

From multiplication to division

Fill in the following ☐ with a number.

2 × ♛ 18☐ ÷ 9 = ☐2	9 × 2 18☐ ÷ 2 = ☐9
3 × ♛ 27☐ ÷ ♛ = ☐3	♛ × 3 27☐ ÷ 3 = ☐9
4 × 9 36☐ ÷ ♛ = ☐4	9 × 4 36☐ ÷ 4 = ☐9
5 × ♛ 45☐ ÷ 9 = ☐5	9 × 5 45☐ ÷ 5 = ☐9
6 × 9 54☐ ÷ ♛ = ☐6	♛ × 6 54☐ ÷ 6 = ☐9
7 × 9 63☐ ÷ 9 = ☐7	♛ × 7 63☐ ÷ 7 = ☐9

Ho Math Chess　何数棋谜　妈！我会棋谜式乘法啦！
Mom! I Learn Multiplication Using Math-Chess-Puzzles Connection!

Student's Name _____ Date _____

2007 - 2017 © Frank Ho, Amanda Ho, All rights reserved.　　www.homathchess.com

From multiplication to division

Fill in the following ☐ with a number.

$\begin{array}{r} 2 \\ \times\ 8 \\ \hline \end{array}$ $16\square \div 2 = \square 8$	$\begin{array}{r} 6 \\ \times\ 2 \\ \hline \end{array}$ $12\square \div 2 = \square 6$
$\begin{array}{r} ♘ \\ \times\ 7 \\ \hline \end{array}$ $21\square \div 3 = \square 7$	$\begin{array}{r} ♖ \\ \times\ 3 \\ \hline \end{array}$ $15\square \div 3 = \square 5$
$\begin{array}{r} 4 \\ \times\ 6 \\ \hline \end{array}$ $24\square \div 6 = \square 4$	$\begin{array}{r} 4 \\ \times\ 4 \\ \hline \end{array}$ $16\square \div 4 = \square 4$
$\begin{array}{r} ♖ \\ \times\ 5 \\ \hline \end{array}$ $25\square \div 5 = \square 5$	$\begin{array}{r} 7 \\ \times\ 5 \\ \hline \end{array}$ $35\square \div ♖ = \square 7$
$\begin{array}{r} 6 \\ \times\ 4 \\ \hline \end{array}$ $24\square \div 4 = \square 6$	$\begin{array}{r} 8 \\ \times\ 6 \\ \hline \end{array}$ $48\square \div 8 = \square 6$
$\begin{array}{r} 7 \\ \times\ 8 \\ \hline \end{array}$ $56\square \div 7 = \square 8$	$\begin{array}{r} 6 \\ \times\ 7 \\ \hline \end{array}$ $42\square \div 6 = \square 7$

Ho Math Chess 何数棋谜 妈！我会棋谜式乘法啦！
Mom! I Learn Multiplication Using Math-Chess-Puzzles Connection!

Student's Name _____ Date _____

2007 - 2017 © Frank Ho, Amanda Ho, All rights reserved.   www.homathchess.com

From multiplication to division

Fill in the following ☐ with a number.

$\begin{array}{r} 2 \\ \times\ 8 \\ \hline \end{array}$ $16\square \div 2 = \square 8$	$\begin{array}{r} 8 \\ \times\ 2 \\ \hline \end{array}$ $16\square \div 8 = \square 2$
$\begin{array}{r} 4 \\ \times\ 9 \\ \hline \end{array}$ $36\square \div ♛ = \square 4$	$\begin{array}{r} 4 \\ \times\ 3 \\ \hline \end{array}$ $12\square \div ♞ = \square 4$
$\begin{array}{r} 4 \\ \times\ 6 \\ \hline \end{array}$ $24\square \div 4 = \square 6$	$\begin{array}{r} 8 \\ \times\ 4 \\ \hline \end{array}$ $32\square \div 4 = \square 8$
$\begin{array}{r} 5 \\ \times\ 7 \\ \hline \end{array}$ $35\square \div 7 = \square 5$	$\begin{array}{r} 4 \\ \times\ 5 \\ \hline \end{array}$ $20\square \div 4 = \square 5$
$\begin{array}{r} 6 \\ \times\ 8 \\ \hline \end{array}$ $48\square \div 6 = \square 8$	$\begin{array}{r} 6 \\ \times\ 6 \\ \hline \end{array}$ $36\square \div 6 = \square 6$
$\begin{array}{r} 7 \\ \times\ ♞ \\ \hline \end{array}$ $21\square \div 3 = \square 7$	$\begin{array}{r} 4 \\ \times\ 7 \\ \hline \end{array}$ $28\square \div 7 = \square 4$

From multiplication to division

Fill in the following ☐ with a number.

☐27　　☐27 　　　÷　3 × 3　　　9 ♛	☐30　　☐30 　　　÷　5 × ♜　　　6 6
☐45　　☐45 　　　÷　5 × 5　　　♛ 9	☐42　　☐42 　　　÷　7 × 7　　　6 6
☐35　　☐35 　　　÷　♜ × 5　　　7 7	☐72　　☐72 　　　÷　8 × 8　　　9 ♛

From multiplication to division

Fill in the following ☐ with a number.

☐45   ☐45
———   ÷ ♖
× 5   ♕
  9

☐48   ☐48
———   ÷ 8
× 8   6
  6

☐40   ☐40
———   ÷ 5
× ♖   8
  8

☐24   ☐24
———   ÷ 4
× 4   6
  6

☐42   ☐42
———   ÷ 6
× 6   7
  7

☐24   ☐24
———   ÷ 8
× 8   3
  ♗

Ho Math Chess 何数棋谜 妈！我会棋谜式乘法啦！
Mom! I Learn Multiplication Using Math-Chess-Puzzles Connection!

Student's Name _____ Date _____

2007 - 2017 © Frank Ho, Amanda Ho, All rights reserved.   www.homathchess.com

From multiplication to division

Fill in the following ☐ with a number.

Ho Math Chess 何数棋谜 妈！我会棋谜式乘法啦！
Mom! I Learn Multiplication Using Math-Chess-Puzzles Connection!
Student's Name _____ Date _____

2007 - 2017 © Frank Ho, Amanda Ho, All rights reserved.   www.homathchess.com

From multiplication to division

Fill in the following ☐ with a number.

☐15　　☐15 $\div \dfrac{5}{3}$ $\times \dfrac{\text{♖}}{3}$	☐27　　☐27 $\div \dfrac{\text{♘}}{9}$ $\times 3$ ♛
☐18　　☐18 $\div \dfrac{6}{\text{♘}}$ $\times 6$ ♘	☐24　　☐24 $\div \dfrac{8}{3}$ $\times 3$ 8
☐21　　☐21 $\div \dfrac{3}{7}$ $\times 7$ 3	☐12　　☐12 $\div \dfrac{4}{3}$ $\times 4$ ♗

2007 - 2017 © Frank Ho, Amanda Ho, All rights reserved.　　www.homathchess.com

From multiplication to division

Fill in the following ☐ with a number.

☐30　☐30 ——— ÷ ♖ × 5　6 6	☐54　☐54 ——— ÷ 6 × 6　♛ 9
☐48　☐48 ——— ÷ 6 × 6　8 8	☐56　☐56 ——— ÷ 7 × 7　8 8
☐42　☐42 ——— ÷ 7 × 6　6 7	☐24　☐24 ——— ÷ 6 × 6　4 4

Ho Math Chess   何数棋谜   妈！我会棋谜式乘法啦！
Mom! I Learn Multiplication Using Math-Chess-Puzzles Connection!

Student's Name _____ Date _____

2007 - 2017 © Frank Ho, Amanda Ho, All rights reserved.   www.homathchess.com

介紹何数棋谜

何数棋谜=奧数棋谜 + 思唯腦力開發
英文教材, 中英双语教学

什麼是何数棋谜?

上百篇科學論文已發表國際象棋可以提高兒童問題解答能力. 並且訓練他們的專心及耐力. 所以我們已經知道下國際象棋對兒童有好處. 但是因為國際象棋與計算能力並無直接開係,所以如何讓兒童能在一個歡樂的環境下也能利用下棋來提高數學的計算呢? 何老師首創並發明有版权的幾何棋藝符號並利用此符號發明了世界第一的独特结合數學与棋谜教材. 何数棋谜讓兒童能利用幾何棋藝符號進行邏輯推理及數字的運算. 棋藝與算術的綜合題含蓋了整數,幾何,集合,抽象數,對比異同,函數,座標,多空間圖形資料,及規則性數字分析. 並且把棋藝的趣味性和數學的知識性結合在一起.

何数棋谜如何幫助兒童腦力思唯的開發?

很簡單的一個道理就是讓學生自願地去用腦,何数棋谜首創獨一無二的融合數學與棋谜的独特趣味寓教於樂教材,利用國際象棋訓練右腦的座標,空間分析及圖形處理,並利用發明了整合棋子與數學的圖形語言,讓兒童能利用符號圖形訓練左腦進行邏輯推理及數字的運算. 國際象棋與算術的綜合題含蓋了整數,幾何,集合,抽象數,對比異同,函數,多空間圖形資料. 所以枯燥無味的計算題變成了謎題,學生需要通過更多的思考. 能讓腦去思考愈多則腦力也愈開發. 處里訊息,分析資料才能發掘出題目. 做這些謎題式數學時可以训練學生比較會專心及有耐心.

Ho Math Chess   何數棋謎   妈!我会棋谜式乘法啦!
Mom! I Learn Multiplication Using Math-Chess-Puzzles Connection!

Student's Name _____ Date _____

2007 - 2017 © Frank Ho, Amanda Ho, All rights reserved.   www.homathchess.com

何數棋謎融合數學與國際象棋的教學理論已在 BC 省數學教師刊物上發表. 科研報告已經證實何數棋謎教學法不但可以提高兒童數學解題及思維能力,還可以開發兒童的腦力,及分析問題的能力並且增加兒童學習的耐力,學生的探索創造精神及求知欲. 判斷力,及自信心等,啓發思維訓練機警靈巧及加強手腦眼的靈活運用.

2007 - 2017 © Frank Ho, Amanda Ho, All rights reserved.   www.homathchess.com

Introducing Ho Math Chess™

Ho Math Chess™= math + puzzles + chess

Frank Ho, a Canadian math teacher, intrigued by the relationships between math and chess after teaching his son chess started **Ho Math Chess™** in 1995. His long-term devotion of research has led his son to become a FIDE chess master and Frank's publications of over 20 math workbooks. Today **Ho Math Chess™** is the world largest and the only franchised scholastic math, chess and puzzles specialty learning center with worldwide locations. **Ho Math Chess™** is a leading research organization in the field of math, chess, and puzzles integrated teaching methodology.

There are hundreds of articles already published showing chess benefits children and that math puzzles are a very good way of improving brainpower. So, by integrating chess and mathematical chess puzzles together, the learning effect is more significant.

Parents send their children to **Ho Math Chess™** because they like **Ho Math Chess™** teaching philosophy – offering children problem-solving questions in a variety of formats. The questions could be pure chess, chess puzzles or mathematical chess puzzles in nature of logic, pattern, tree structure, Venn diagram, probability and many more math concepts.

Ho Math Chess™ has developed a series of unique and high quality math, chess, and puzzles integrated workbooks. **Ho Math Chess™** produced the world's first workbook **Learning Chess to Improve Math.** This workbook is not only for learning chess, but also for enriching math ability. This sets **Ho Math Chess** apart from other math learning centers, chess club, or chess classes.

The teaching method at **Ho Math Chess™** is to use math, chess, and puzzles integrated workbooks to teach children fun math. The purposes of **Ho Math Chess™** teaching method and workbooks are to:

- Improve math marks.
- Develop problem solving and critical thinking skills.
- Improve logic thinking ability.
- Boost brainpower.

Testimonials, sample worksheets, reports, and franchise information can be found at www.homathchess.com.

More information about **Ho Math Chess™** can also be found from the following publications:

1. Why Buy a **Ho Math Chess™** Learning Centre Franchise: A Unique Learning Centre?
2. **Ho Math Chess™** Sudoku Puzzles Sample Worksheets
3. Introduction to **Ho Math Chess™** and its Founder Frank Ho

The above publications can be purchased from www.amazon.com.

www.ingramcontent.com/pod-product-compliance
Lightning Source LLC
Chambersburg PA
CBHW082126210326
41599CB00031B/5891

9 781927 814888